Ocean Treasure
Commercial Fishing in Alaska

by **Terry Johnson**
edited by **Kurt Byers**

Alaska Sea Grant College Program

Elmer E. Rasmuson Library Cataloging in Publication Data

Johnson, Terry Lee, 1947-
Ocean treasure : commercial fishing in Alaska / by Terry Johnson ; edited by Kurt Byers. – Fairbanks, Alaska : Alaska Sea Grant College Program ; University of Alaska Fairbanks, 2003.
200 p. : ill. ; cm.

Note: "NOAA National Sea Grant Office, grant no. NA16RG-2321, projects A/161-01 and A/151-01."

Includes bibliographical references.

ISBN 1-56612-080-2

1. Fisheries—Alaska. 2. Fishery innovations—Alaska. 3. Fishery technology—Alaska. 4. Fishery management—Alaska. I. Title. II. Johnson, Terry Lee. III. Byers, Kurt.

SH214.4.J64 2003

Front cover photo/illustrations: top left to right: ©2003, Mark Emery, AlaskaStock.com; Bob Hitz; Kurt Byers; Mandy Merklein. Bottom: all Tony Lara. Back cover: Robert Lauth; Rick Harbo; Bob Hitz; John Hyde, ADFG; Victoria O'Connell. Design by Tatiana Piatanova, Alaska Sea Grant.

Credits:

This book is published by the Alaska Sea Grant College Program, which is cooperatively supported by the U.S. Department of Commerce, NOAA National Sea Grant College Program, grant no. NA16RG-2321, projects A/161-01 and A/151-01; and by the University of Alaska Fairbanks with state funds. University of Alaska is an affirmative action/equal opportunity institution. The views expressed herein do not necessarily reflect the views of the above organizations.

Sea Grant is a unique partnership with public and private sectors combining research, education, and technology transfer for public service. This national network of universities meets changing environmental and economic needs of people in our coastal, ocean, and Great lakes regions.

Alaska Sea Grant College Program
University of Alaska Fairbanks
P.O. Box 755040
Fairbanks, Alaska 99775-5040
Toll free (888) 789-0090
(907) 474-6707 Fax (907) 474-6285
http://www.uaf.edu/seagrant/

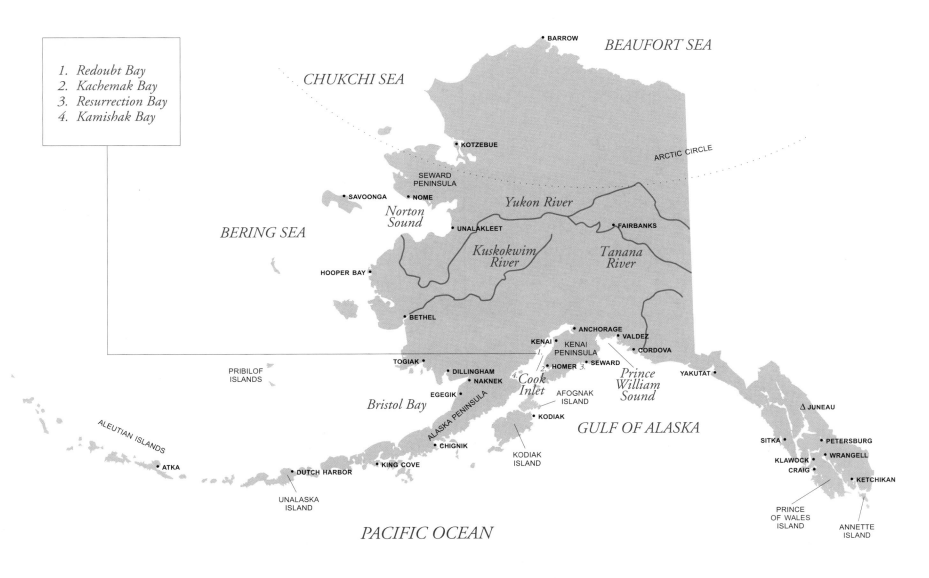

ARCTIC OCEAN

BEAUFORT SEA

CHUKCHI SEA

• BARROW

1. Redoubt Bay
2. Kachemak Bay
3. Resurrection Bay
4. Kamishak Bay

ARCTIC CIRCLE

• KOTZEBUE

SEWARD
PENINSULA

• SAVOONGA

Yukon River

• NOME

Norton
Sound

BERING SEA

• UNALAKLEET

• FAIRBANKS

Kuskokwim
River

Tanana
River

HOOPER BAY •

• BETHEL

• ANCHORAGE   • VALDEZ

KENAI •   Kenai   • CORDOVA
        Peninsula
1.   • SEWARD
TOGIAK •   2. • HOMER 3.   Prince
                                   William   • YAKUTAT
PRIBILOF   • DILLINGHAM 4. Cook   Sound
ISLANDS   • NAKNEK   Inlet
                              AFOGNAK
        EGEGIK •   ISLAND   △ JUNEAU
Bristol Bay   ALASKA PENINSULA
                            • KODIAK
ALEUTIAN ISLANDS   GULF OF ALASKA   SITKA •   • PETERSBURG

                • CHIGNIK   KODIAK   KLAWOCK • • WRANGELL
• ATKA   • DUTCH HARBOR   • KING COVE   ISLAND   CRAIG •
                                                                  • KETCHIKAN
        UNALASKA
        ISLAND   PRINCE
                                                  OF WALES
                                                  ISLAND   ANNETTE
PACIFIC OCEAN   ISLAND

Troller poles accent the harbor at Wrangell, Alaska.

Contents

Foreword

Senator Ted Stevens addresses Sea Grant staff and supporters at Sea Grant's 30th anniversary reception at the U.S. Capitol in Washington, D.C.

Long before Alaska was a United States territory, our people were fishing for their livelihood. Fish were essential in the lives of Native Alaskans, providing a valuable food source and needed materials. After the Russians came to Alaskan soil they built salteries to preserve their catch so that they would have enough food to get them through the harsh winter. Then, in 1878, the first canneries were established at Sitka and Klawock, marking the beginning of our commercial fishing industry. For Alaskans, fishing is more than just an industry; it is intertwined with our way of life.

Today, that tradition continues. Our waters are home to large numbers of salmon and other fish species. Native Alaskans have relied upon this resource for thousands of years and fishing continues to be a key part of our culture and commitment to subsistence living.

Fishing also represents an enormous piece of the Alaskan economy. Every year thousands of recreational fishermen take to Alaska's waterways and another generation learns about this integral part of our heritage. To truly understand the impact of the commercial fishing industry in Alaska you need only to look to the great people of our state. Thousands of Alaskans derive income or benefits from the commercial fishing industry. We understand better than anyone the importance of sustainable fisheries and a healthy and vibrant ecosystem in the North Pacific.

Personally, I am committed to preserving this essential part of our culture. As an original author of the Magnuson-Stevens Fishery Conservation and Management Act, I am dedicated to exploring measures that will improve the management of our fisheries. My interest in this industry extends beyond policy; as a recreational fisherman and a father of a former commercial fisherman, my goal is to ensure that this resource is available to future generations. Fishing is Alaska's greatest pastime and this book aims to preserve this part of our heritage.

Terry Johnson has depicted well the fishing industry's importance to our state and our nation. While he looks back to uncover the history of fishing in our state, this book also looks forward to some of our key challenges. My hope is that it will educate readers, Native Alaskan and newcomer alike, about fishing in our state and the brave Alaskans who take to the sea to keep this industry thriving.

Ted Stevens

The Honorable Ted Stevens
United States Senate

Preface

Alaska's seafood harvesting and processing industry is the state's second largest private industry and is the foundation for much of Alaska's social structure. Without productive marine and aquatic ecosystems, this and many other ocean-dependent industries and cultures in Alaska cannot survive.

For Alaska's marine resources to remain robust, we must understand how our actions combine with natural factors to affect the environment. Then we should prudently apply our knowledge as we interact with the marine and coastal environments in ways that will not irreversibly diminish them.

A well-informed public is one of the keys to keeping the globe's marine resources in good shape into the foreseeable future. This book is the final component in a three-part public awareness project begun by Alaska Sea Grant in 1991 to help Alaskans and non-Alaskans alike better understand Alaska's commercial fishing industry (see Anatomy of a Public Information Project in the back of this book). Our intent is to provide easy-to-digest, authoritative information about the history, fishing techniques, species harvested, and management of this colorful, complicated, controversial, and fascinating business.

The fishing industry and the regulatory and natural environments in which it operates are constantly in flux. One of the challenges in creating this book was to include information that accurately describes the industry at the beginning of the twenty-first century without including a lot of detail that may become outdated or irrelevant soon after publication. Still, in order to make essential points about key aspects of the industry, we included some information—such as ex-vessel prices, harvest totals, and other content—that will likely be outdated soon after this book is published. Readers seeking current statistical data on the industry should contact the Commercial Fisheries Division of the Alaska Department of Fish and Game or the National Marine Fisheries Service.

Throughout this book, we point out some of the most critical challenges facing the industry and describe how fishermen, policy makers, resource managers, and other concerned parties are addressing the problems. While we expect this book to have a long useful life, we hope the problems described will not share the same longevity.

Our ultimate goal is that this publication will lead to a better understanding of the Alaska commercial fishing industry and the role it plays in the state, national, and international communities. If this document helps inform a constructive dialogue among people who are waging the same battle on often different fronts to maintain the vitality and integrity of the North Pacific ecosystem, then it will have done its job.

—Kurt Byers
Editor

Acknowledgments

DAVE BRENNER

Terry Johnson and Kurt Byers look over a few transparencies for use in this book.

The primary author of this publication is Terry Johnson, University of Alaska Fairbanks (UAF) associate professor of fisheries and Marine Advisory Program (MAP) agent in Homer, Alaska. Johnson fished commercially in Alaska from 1978 to 1995, in Southeast Alaska, Yakutat, Bristol Bay, Yukon River, and Norton Sound. He has written another book about the Alaska commercial fishing industry, titled *Alaska Fisheries Handbook*, published by Seatic Publishing in Sitka, Alaska. Johnson is a regular contributor to the commercial fishing trade press and has authored several Alaska Sea Grant publications about the business of commercial fishing.

Johnson is an advisor to the Alaska commissioners of the Pacific States Marine Fisheries Commission. He also runs an ecotour charter business in Bristol Bay, Alaska, and is a charter member of the Alaska Marine Conservation Council board of directors. Johnson, a U.S. Marine Corps veteran, holds a B.A. in communications and an M.A. in marine resource management, both from the University of Washington, Seattle.

Contributing authors include Mike Banks, Kurt Byers, Paula Cullenberg, Charlie Ess, Lesley Leyland Fields, Dolly Garza, Tony Lara, Rich Mattson, Ray RaLonde, and Doug Schneider.

Mike Banks is a dive harvest and Tanner crab fisherman in Petersburg, Alaska. He contributed information on dive harvest techniques.

Kurt Byers is communications manager, writer/editor, and photographer with the Alaska Sea Grant College Pprogram (ASG). He conceived, compiled, and edited the book, assembled and selected photographs and information graphics; wrote photo captions and the sidebars on the oil spill, bycatch, seafood pricing, underwater deforestation, and the "humpy dump."

Paula Cullenberg is former director of the North Pacific Fisheries Observer Training Center at the University of Alaska Anchorage (UAA) and current MAP leader in Anchorage. A former groundfish observer and setnet and driftnet fisherman, she wrote the sidebar about the federal observer program.

Charlie Ess is a writer and photographer based in Wasilla, Alaska who specializes in the Alaskan commercial fishing industry. He helped with the jigging and clamming sections.

Lesley Leyland Fields is professor and chair of the English Department at UAA on Kodiak Island. She is a setnet fisherman and author of *The Entangling Net*, a book about women in Alaska fisheries, and *The Water Under Fish*, a book of poetry about life with Alaska seas.

Dolly Garza is UAF professor of fisheries and MAP agent in Ketchikan, Alaska. A Tlingit/Haida Indian, Garza contributed to the subsistence chapter.

Tony Lara is a former Bering Sea crab and longline fisherman and photographer who lives in Kodiak, Alaska. He reviewed the text and wrote parts of the longline and crab fisheries sections, and his photos appear throughout the book.

Rich Mattson is a fisheries biologist and director of the Douglas Island Pink and Chum Hatchery in Juneau, Alaska. A non-formal fisheries historian, he provided the information about fish wheels.

Ray RaLonde is UAF professor of fisheries and MAP aquaculture specialist, located in Anchorage. RaLonde wrote most of the section on shellfish mariculture.

Doug Schneider is public information officer and producer of the Arctic Science Journeys radio program at ASG. He contributed the information in the sidebar on electronic technology in commercial fishing and wrote the sidebar on fish stock assessment.

We also gratefully acknowledge critical review of the text provided by many experts in various fields of the Alaska seafood industry, management bodies, and conservation groups. For review of the original manuscript, our thanks go to Bob King, special assistant to former Alaska Governor Tony Knowles, journalist, and commercial fisheries historian; Robert Larsen, herring and dive fisheries project leader with the Alaska Department of Fish and Game (ADFG) in Petersburg; Denby Lloyd, ADFG Westward Region supervisor; Brad Matsen, author and past West Coast editor of *National Fisherman*; Bob Mikol, a commercial fisherman, author of the book *Temperature Directed Fishing*, published by ASG, and formerly a member of the board of directors of the Alaska Marine Conservation Council; Doug Schneider, ASG; Leigh Selig, enforcement officer with the National Marine Fisheries Service (NMFS); and John van Amerongen, editor of *Alaska Fisherman's Journal*.

UAF scientists Gerry Plumley and John French reviewed the information on paralytic shellfish poisoning. Terrance Quinn, former co-chair of the Scientific and Statistical Committee of the North Pacific Fishery Management Council and UAF professor of fisheries, reviewed the sidebar on bycatch. Stephen Jewett and Brenda Konar, scientists with UAF, contributed information about the life history of crabs (Jewett) and kelp forest ecosystem dynamics (Jewett and Konar).

Gunnar Knapp, UAA Institute of Social and Economic Research, and Chris McDowell of the McDowell Group in Juneau, provided information on salmon pricing and distribution. Tim Ryan of Sitka Sound Seafoods also helped with the salmon pricing section. Chuck Crapo, MAP seafood quality specialist, reviewed the chapter on processing and marketing. Rodger Painter, executive director of the Alaska Shellfish Growers Association, reviewed text on shellfish mariculture. Al Burch, president of the Alaska Draggers Association, and Steve Patterson of New England Trawl Systems, provided information about trawling techniques. Jerry Dzugan, executive director of the Alaska Marine Safety Education Association, and Greg Switlik, president of Switlik Parachute Co., reviewed the information on marine safety and survival. Guy Hoppen, tender operator, reviewed and provided information on purse seining and tendering.

A.J. Paul, crab biologist and UAF professor emeritus, provided information about crab nomenclature and life history and reviewed the accuracy of the crab illustrations. Robert Larsen, ADFG, provided information about sea urchins, sea cucumbers, and geoducks. Tim Koeneman and Michael Ruccio, ADFG, provided catch statistics and other information about shrimp and crab fishing. Victoria O'Connell, Jeff Barnhart, Doug Woodby, and Jim Blackburn, all of ADFG, contributed range and other information on shellfish and finfishes. Joan Forsberg, biologist with the International Pacific Halibut Commission (IPHC), provided range information for halibut, and Heather Gilroy, biologist with IPHC, provided catch statistics. Thomas Kong, biologist with IPHC, helped locate halibut photos. Andrew Trites, University of British Columbia, and Kate Wynne, MAP marine mammal specialist, reviewed the information on Steller sea lion population change.

A special thanks goes to Herman Savikko, statistician with ADFG, for his manuscript review and assistance providing fisheries catch statistics. Savikko was extraordinarily responsive to many requests for statistics, text review, and other information.

David Sutherland, fishery biologist with NMFS, Fishery Statistics and Economics Division, provided statistics on five-year average catch volumes (1994-1998) found in the Sources and Useful References chapter. Mike Plotnick, research analyst with ADFG, provided statistics on harvest value. Elizabeth Logerwell, biologist at the NOAA Alaska Fisheries Science Center in Seattle, helped locate groundfish photographs. Tracey Lignau, biologist with ADFG, provided information about fishing activity in Hooper Bay.

Thanks also to the honorable U.S. Senator Ted Stevens and his staff for providing the foreword to this publication.

Pre-press review of the book was provided by Jim Branson, former director of the North Pacific Fishery Management Council, Brian Paust, UAF professor emeritus and former MAP agent in Petersburg, and Paula Cullenberg, MAP.

Thanks also to Dave Gordon, Washington Sea Grant; and the late Richard Carlson, NMFS, for tips on finding photographers whose work appears in this book. Thanks to Angela Linn, anthropologist with the University of Alaska Museum, for directing the editor to Native artifacts in the museum's collection, photos of which appear in the Subsistence Fishing chapter.

Book design, cover, and information graphics are by Tatiana Piatanova, ASG graphics manager, assisted by Kurt Byers. Gear illustrations are by Bob Hitz. Pen-and-ink illustrations of fish and shellfish are by Sandra Noel and Lisa Peñalver. Colorization of the drawings is by Lisa Peñalver. Sue Keller, ASG publications manager, proofread the text, oversaw printing, and provided editorial advice. The index was done by Nanette Cardon.

This book is dedicated to the scientists, government staff, economists, lawyers, administrators, activists, and politicians who, somehow, collectively work together to keep Alaska fisheries healthy, sustainable, and productive.

Terry Johnson

INTRODUCTION

Commercial fishing in Alaska is a diverse, colorful, tough, dangerous, thriving, and—with skill and luck—lucrative enterprise. This bounty makes the 49th state with its 34,000 miles of shoreline the most important fishing state in the nation.

Alaska's nearshore and offshore waters produce about half the U.S. seafood harvest every year, with an average dockside value of about $1.4 billion. Not only does the Alaska seafood harvest outrank in quantity and value the harvest of the rest of the United States combined, this bounty also outranks the individual harvests of Norway, Denmark, Iceland, and Canada—each a well-known fishing nation.

Dutch Harbor and Kodiak usually rank among the top five fishing ports in the country. In recent years, the huge offshore groundfish fishery has elevated Dutch Harbor in the Aleutian Islands to the nation's number one port city in quantity landed.

The seafood industry has long been Alaska's largest private industry employer, currently providing full- and part-time jobs for some 70,000 men and women. And unlike industries that depend on nonrenewable resources, the seafood industry has excellent long-term prospects. Given the massive volume and value of Alaska fisheries and the prospect that the resource, if managed well, should be around for a long time, seafood harvesting and processing represents one of Alaska's best hopes for long-term economic health.

This book will give you an overview of the commercially important fish and shellfish that thrive in Alaska waters, and insight into the tools, techniques, and management structure used to harvest and maintain Alaska's vast ocean bounty.

The Great Divide

Although there is some overlap in species harvested and gear types used in nearshore and offshore Alaska waters, the terms nearshore and offshore generally distinguish the two fundamental segments of Alaska's $1.4 billion (dockside value) fishing industry.

The offshore fishing industry, which operates in federally managed waters from three to 200 miles from shore, is composed primarily of large crabbers, longliners, and trawlers based in southcentral Alaska or Washington state. They harvest crab, pollock, cod, and other groundfish, which account for just over half the total annual value of the Alaska harvest.

Alaska's nearshore fishery is conducted mostly by smaller operators who fish primarily in waters managed by the State of Alaska, which extend three miles out from shore. These fishermen harvest salmon, herring, halibut, blackcod, crab, and shrimp, among other species.

Alaska Fishery Fuels Local, State, and National Economies

Alaska's commercial fishing industry has four key elements: harvesting, processing, support services, and marketing. Approximately 40,000 men and women help harvest Alaska's fish and shellfish each season, mostly on board the nearly 18,000 fishing boats which ply waters off Alaska shores. Another 30,000 people work in processing, mainly in cannery and cold storage facilities. The number of workers in the service sector is difficult to estimate, but includes people who work in fuel delivery, vessel repair, air freight, food service, law, insurance, banking, and other occupations. The seasonal nature of the industry makes it convenient summer employment for college students.

Adventurous college students can find unique summer employment in Alaska's seafood processing plants, such as these women who are helping offload halibut from a longline vessel in Homer.

Most Alaska seafood is exported, with Japan the top destination. These exports are an important factor in reducing America's trade deficit. Much of the revenue generated by the seafood industry is recycled through the state economy in the form of goods and services purchased from local vendors, raw-fish taxes, local sales taxes, property taxes on equipment and facilities, and personal income spent within the state. However, because about forty percent of fishing vessel crew members are non-residents, and because the majority of the processing capacity is owned by non-resident and foreign

Halibut longliners take a well-earned breather after offloading their catch at a Cordova processing plant.

corporations, a large portion of Alaska's seafood industry revenue is exported from the state.

Almost all of Alaska's 600,000 citizens live near the sea or a major river. This makes fishing—and the industries that support fishing—a key source of income for the great majority of Alaska communities, from Ketchikan at the southern end of the panhandle to Kotzebue above the Arctic Circle.

Anchorage, Alaska's biggest city, has the state's largest population of commercial fishermen, even though the city is not home to a resident fishing fleet and does not have facilities to receive direct delivery of fish from vessels. Fairbanks, located in the central interior of Alaska more than 350 miles from the nearest saltwater, counts scores of commercial fishermen among its residents. And people in many other communities far inland along the Yukon and Kuskokwim rivers, in the Bristol Bay region, and elsewhere heavily depend on fish for food and income.

Fishermen Are Versatile and Mobile

Alaska fishermen tend to be mobile, diversified, and drawn to the industry by its unique appeal. It is not uncommon for an Alaska fisherman to participate in three or more fisheries in one year. A fisherman may fish blackcod in April, herring in May, halibut in June, salmon in July and August, crab in the fall, and groundfish in the winter. Others may log an intensive three weeks chasing Bristol Bay sockeyes in July and kick back and relax the rest of the year. Another group of enterprising fishermen may put in short seasons in California or Washington.

A longline fisherman seems pleased with a 350-plus pound halibut, taken off Kodiak Island.

Some fishermen own two or more boats located in different parts of the state, and fly from one to another depending on what fishery is open at a given place and time. Other fishermen own a single boat worth $2 million and spend 40 weeks a year running it. Still others simply fish from open skiffs, pulling their nets and lines by hand, or set and pick their nets on mud flats between tides.

In recent decades many fishermen have earned sizable incomes. The lure of fast money has drawn many people from other professions. Fleets have former attorneys, physicians, psychologists, professors, secretaries, airline pilots, teachers, homemakers, salesmen, and politicians who have responded to the call of adventure, big bucks, and independence. Many

GUY HOPPEN

Purse seine crew from the fishing vessel
Mermaid II takes a break.

have permanently left their former occupations. Others have found they can put in a short season and return to their regular jobs. Still others test the waters for a year or more, decide fishing isn't for them, and return to their former lives with an experience of a lifetime.

The dawn of the new century, however, has seen a downturn in the fortunes of some of Alaska's fisheries. Diminished stocks are the cause in a few cases, such as king crab, while market competition or reduced consumer demand which is driving down prices paid to fishermen is the culprit in others, like salmon and herring. Lower financial returns are driving some fishermen out of the business and forcing others to find supplementary income.

DAVID F. KAPLAN

Salmon setnetting and drift gillnetting sometimes involve whole families. These kids help out in the family setnet business in Kodiak.

Alaska's Fishing Women: From Dock to Deck

These days, male and female fishermen routinely work together on fishing vessels. Here a woman and a man haul in a nice sockeye salmon aboard a sternpicker in Bristol Bay. Nearly all female fishermen who responded to a survey conducted in the 1990s by the Kodiak, Alaska, newspaper said that they want to be referred to as "fishermen," not "fisher" or "fisherwoman" or any other term.

The Alaska commercial fishing industry is populated mostly by men, but women work in every part of the industry. No one knows for sure, but a good estimate is that women make up about five percent of the workforce, not including women in seafood processing plants.

Historically, Native women were an active part of commercial fishing. Indeed, in some regions and fisheries, such as the Yukon River and Bristol Bay, commercial and subsistence setnet fishing was considered women's work. However, as the industry modernized and became intensely competitive, Native women became less visible in commercial fishing, and now men handle most of the commercial fishing chores in Native communities.

In the non-Native world, as women began breaking gender barriers in other professions in the 1970s, a wave of adventure-seeking women migrated north and worked their way into what had become an almost exclusively male occupation.

For some the transition was eased by partnerships with boyfriends and husbands. For others who traveled and worked alone, the step from dock to deck was difficult, sometimes even dangerous. Men who resisted women's sudden presence greeted the women with varying degrees of discrimination and harassment, while other fishermen were supportive and encouraging.

Many female Alaska fishermen got their start by mending nets and building crab pots between countless hours walking the docks trying to coax, cajole, and otherwise convince male fishermen they could do the job. Today women are often seen setting seines, hauling and picking gillnets, and doing everything else necessary to succeed on the fishing grounds.

Beach seiners are geared up for a salmon run somewhere in Alaska around 1890.

A SHORT HISTORY

KURT BYERS

For millennia prior to the arrival of the first Europeans, Native Indians, Eskimos, and Aleuts relied on fish as a prime barter commodity in their subsistence culture. When Russian hunters and traders first arrived in the late 1770s in search of sea otter pelts, Natives supplied them with food, including the abundant salmon, halibut, and cod.

The Russians quickly recognized the fishery potential and built fish processing facilities, primarily to feed their garrisons. Russians salted fish at Sitka and at nearby Redoubt Bay in the early 1800s.

Fish canning technology was developed in California and soon spread north. Alaska's first salmon cannery opened at Klawock in Southeast Alaska in 1878. Within five years the industry spread northwest all the way to Bristol Bay. By 1900 it seemed that nearly every bay and major river mouth had at least one cannery, herring saltery, or salmon curing operation. Plants were built to render herring to oil and meal, and even whaling stations appeared along the Gulf of Alaska coast. The "great Alaskan fish rush" was on.

Although the fish rush never got the popular attention the gold rush did, the annual value of Alaska's canned salmon alone often exceeded the territory's yearly production of gold.

But the party did not last forever. After decades of unbridled exploitation, it became clear that Alaska fish stocks were not limitless. By the 1940s salmon production was in a downward spiral.

Enlightened Management Saves the Fishery

Many people blamed the salmon decline on highly efficient fish traps. Alaskans saw the devices as symbolic of oppression by out-of-territory economic interests because they gave the canneries virtual ownership of the salmon resource. Opposition to salmon traps was so intense that it became one of the driving forces behind the statehood movement. When Alaska became a state in 1959, fish traps were immediately banned by the new state legislature.

A diver prepares to unclog a chicken wire salmon trap at False Pass, circa 1950.

But just banning traps was not enough to save the salmon. State managers also took an aggressive approach to monitoring and regulating the fishery. Over time a system of managing fisheries was developed based on the concepts of "harvestable surplus" and "maximum sustainable yield." This approach meant that fishermen would be allowed to catch only those fish which biologists determined were surplus to the total number of fish necessary to sustain the fish population at its most productive level.

This was accomplished, in part, by limiting fishermen's gear, fishing time and areas, and eventually by limiting the number of fishermen. Commonly known as "limited entry," the system initiated by the Alaska legislature in 1974 limits the number of fishing permits for salmon, herring, and other fisheries. Permits were originally awarded to fishermen based on their past participation and economic dependence on the fishery. The law allows the sale and trade of permits, which has become an industry in itself. During the peak value years of the 1980s and early 1990s, individual permit values in some fisheries were as high as $400,000.

Limited entry—combined with hatchery production, habitat enhancement, elimination of certain offshore fisheries, and the assistance of favorable environmental conditions—has helped Alaska's salmon resource bounce back to record levels of production.

As of 2000, Alaska had limited entry in some herring, blackcod, and crab fisheries and all salmon fisheries. Federally managed halibut and groundfish fisheries are under individual fishing quotas or other limited access management systems.

People, Nature Affect Fish Stocks

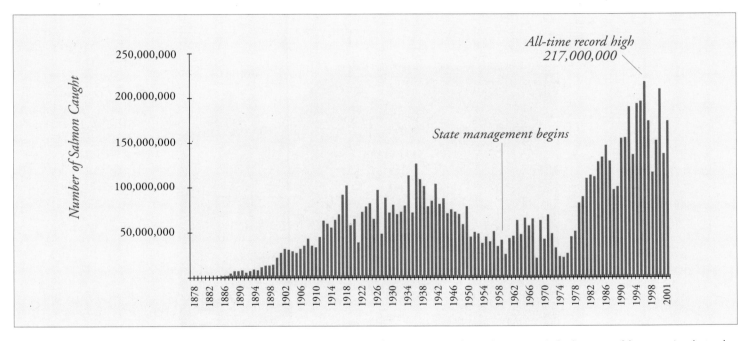

Alaska's fisheries dramatically fluctuate due to human influences and environmental change. Near cataclysmic declines in salmon stocks between World War II and the mid-1970s were mostly due to human action—overharvest. Likewise, the subsequent recovery of salmon primarily was due to human action—ban of fish traps, limited entry to fisheries, enhancement of wild stocks via salmon ranching, and habitat rehabilitation.

By 2002, some salmon stocks began to decline again. This time scientists attributed it to a routine cyclical warming and cooling of the North Pacific Ocean, part of a "regime shift."

Scientists point to a regime shift as a likely explanation for the explosion of walleye pollock in the North Pacific in the 1960s.

Likewise, the extreme drop in the king crab stock in the 1980s may have been mostly a result of a change in water temperature rather than overharvest, although fishing activity probably contributed too.

Scientists have only recently begun to examine how the combination of regime shifts and human activity together affect marine ecosystems. Their discoveries will further aid fishery managers in their quest to make the right calls at the right time.

United States Takes Over Huge Offshore Fishery

While the salmon stock recovery was under way, the U.S. government took an interest in fisheries farther out to sea. Huge vessels from Japan, Korea, Poland, the Soviet Union, and other countries had dominated the waters off Alaska's shores, in some cases right up to three miles from the beach. But in 1976 Congress passed the Fisheries Conservation and Management Act, which gave the United States authority to manage resources between three and 200 miles off our shores. This region, originally dubbed the Fishery Conservation Zone, is now called the Exclusive Economic Zone (EEZ).

This shift changed the face of the U.S. commercial fishing industry. As foreign fleets were phased out of the EEZ, the American fishing industry hustled to gear up and learn how to harvest the vast resources of cod, pollock, crab, flatfish, Atka mackerel, and other offshore species.

Crewmen haul in a catch of Pacific cod off Kodiak. P-cod is one of the offshore fisheries that U.S. fishermen began to harvest in earnest after the Magnuson Act passed. The timing was good, too, because P-cod is one of the fisheries that helped fill the gap when the king crab fishery took a nosedive in the late 1980s.

In the early days of the transition, fishermen mostly focused on the valuable king crab resource. But after a few years, and for reasons scientists still do not entirely understand, crab stocks crashed. To stay afloat financially, fishermen converted their boats to trawl or longline rigs and set sail to harvest Pacific cod, pollock, sole, and other groundfish.

Expanding world demand for these so-called "white fish" products stimulated development of an American fleet of catcher-processors which harvest and process fish. By 1993 there were some 70 big factory trawlers, most based in Seattle, operating off Alaska, some longer than a football field and employing upwards of 100 crew members. Federal legislation passed at the end of the twentieth century reduced the offshore trawl fleet by more than two-thirds.

The remaining trawlers supply much of the fish you eat in fast food restaurants. The catch also is processed into surimi, which is a fish paste used to make artificial crab and other products known in the industry as "seafood analogs."

Catcher-processor ships based in Seattle are
docked at Captains Bay, Unalaska Island.

Technology Advances

While government management has played an important part in the development of Alaska fisheries, technological development has had at least as big a role. Huge stocks of king and Tanner crab were largely ignored by American fishermen until after World War II when freezing technology improved. Mechanical refrigeration, coupled with declining salmon runs, also prompted owners to shut down remote canneries and consolidate operations in central locations where they could be served by tenders carrying fish chilled by flake ice or refrigerated seawater.

Today, large freezing operations and improved air and sea transportation facilitate the processing, storage, and distribution of high-quality frozen fish, products much higher in value than fish that have undergone traditional methods of preservation such as salting and canning. Improved preservation and transportation help satisfy U.S. and foreign demand for highly valued fresh seafood.

Huge container ships like this one at Kodiak call on major Alaska fishing ports to load freezer vans full of seafood processed in those towns, and to deliver empty vans.

Scientists are finding ways to get still more value from the sea by developing new uses for fish and shellfish and their byproducts, which used to be discarded as waste. Nearly 85,000 metric tons of top-quality fish meal was produced in 2002 from fish tissues previously discarded as waste. Much of that is purchased at premium prices by eel farmers in Asian countries. Fertilizers are another product now being produced from what used to be considered fish processing waste.

Fish have long been known to be a healthful food. But researchers are discovering and developing new seafood products from fish and invertebrates that can help bolster human nutrition and health. For example, seafood scientists expect greatly increased production of high quality fish oil containing omega-3 fatty acids, which are known to help prevent heart disease and other ailments. This oil is derived from the liver of white fish (cod and pollock) and from fatty fish (salmon) species.

Legacy of a Disaster

While millions of dollars and thousands of hours were spent trying to clean up the 1989 *Exxon Valdez* oil spill, it was ultimately left to nature to restore the ecosystem, a process that is still under way. Scientists found that some cleanup methods did more harm than good. The widely used method of spraying oiled rocky beaches with hot water to loosen the oil resulted in killing organisms that survived the oil, further delaying ecosystem recovery.

The economy and social structure of most coastal towns in Alaska center on commercial fishing. In Southcentral Alaska people took a hard hit from the Exxon Valdez *oil spill, which occurred in March 1989.*

Nearly every herring and salmon fishery in the region was canceled that summer because of fears the oil would contaminate seafood, which it did to a limited extent. The spill also caused major societal conflicts to fester in communities affected by the spill and caused upheaval in Native communities dependent on sea life.

Since 1989, there has been one spill-related district-wide commercial fishing closure, the Prince William Sound herring fishery. That lucrative fishery was closed in 1993 and has not re-opened due to the collapse of the herring population. Viral disease and a fungus were the probable reasons for the herring population crash, but scientists cannot absolutely attribute those afflictions to oil contamination.

In 1991, Exxon agreed to pay a $25 million criminal fine, $100 million criminal restitution, and a $900 million civil settlement. The agreement stipulated that the $900 million must be used to help recovery of the natural resources and the subsistence, recreational, and commercial fisheries. Court cases also have awarded monetary damages to some fishermen in a class action suit.

Settlement money bought hundreds of miles of anadromous waterways, providing basic protection for spawning and rearing sockeye, chum, pink, coho, and king salmon. Many state and federal research projects were funded aimed at improving the health of commercial fish species and providing new tools for better fisheries management. Native subsistence users were consulted to help scientists gain a first-hand historical perspective on how the marine ecosystem hit by the spill had flourished prior to the spill.

Commercial fishermen are now an integral part of the government's spill response strategy. They proved after the 1989 spill that they have the knowledge, experience, and technical skills critical to marshal an effective response to oil spills. A cadre of fishermen is now trained and ready to help deploy oil containment boom. They also will be consulted when urgent decisions must be made about what places most need to be protected from the onslaught of the next killer oil slick.

Tough Challenges Remain

Alaska's fishing industry will always face complex challenges. Currently, fisheries experts and other concerned people are working hard to find ways to use previously ignored species such as arrowtooth flounder and sea cucumbers; develop more sophisticated and effective resource management strategies; find ways to mediate conflicts among fishermen using different gear types and of different nationalities; and find ways for fishermen to coexist with marine mammals and seabirds.

Some Alaskan commercial fisheries experience conflicts over resource allocation with sport fishermen and subsistence users. Some fisheries unintentionally capture and intentionally throw away non-target fish or shellfish (called bycatch) which lack economic value because they are the wrong size or species (economic discards). And in some fisheries, state and federal regulations require that specific non-target species be discarded when caught (regulatory discards). Reduction of this wasted bycatch is one of the biggest fisheries issues of the new century. Disruption or even destruction of key marine habitats also looms large as a problem in need of rational assessment and resolution.

Commercial fishing is an exciting yet troubled industry. About the only thing to be said for sure about its future is that it will continue to change.

TONY LARA

Floating processors in operation at Bristol Bay.

A crab boat bucks choppy
seas in the Gulf of Alaska.

Gillnetters scramble to get their
share of the Bristol Bay sockeye harvest.

BOATS AND GEAR

Use of tools is one thing that sets people apart from the other animals that share the planet. Commercial fishing is the last vestige of the hunter-gatherer phase of human cultural development. The tools of fishing used today evolved directly from those of our ancestors. Nearly all fish catching devices in Alaska are simply improvements on one of three basic types of devices: hooks, nets, and traps.

Over the years, fishermen have developed increasingly efficient gear designed to harvest specific species. As harvest efficiency has increased, so has the number of commercial fishermen. The combined effect is that today's commercial fishing industry is, in most cases, capable of catching a lot more fish than the stocks can biologically support.

To maintain a long-term supply of seafood, government fishery managers develop regulations to limit both total catch and fishing effort. Regulations dictate what kind and how much gear can be used, establish boundaries of the districts in which fishing can occur, and determine what days (or even hours) fishing will be permitted.

This chapter describes the basic gear types used to harvest the most commercially important fish and shellfish of Alaska. The Fisheries Management chapter outlines the tools managers use to regulate these gear types and maintain the fisheries.

Salmon Fishing Gear

Gillnetting, seining, and trolling are the three primary methods used to harvest salmon in the Alaska commercial fishery. A salmon fisherman may select any of several types of fishing gear, depending on area and species to be harvested and the amount of money available to invest.

Gillnets

The most common type of commercial fishing gear used in Alaska is an entangling net called the drift gillnet. Gillnet fishermen take all species of salmon, although most target sockeye, coho, and chums. These inshore gillnets are highly selective gear which in Alaska rarely catch non-target fish and birds. They must not be confused with miles-long high-seas driftnets which, until they were banned in the early 1990s, were used by some foreign fleets. High seas driftnets were notorious for excessive bycatch of fish, marine mammals, and birds.

A fisherman in Southwest Alaska mends a gillnet. Gillnets are used in nearshore ocean waters and in rivers, primarily to catch herring and salmon.

Gillnets are made of tough, lightweight twine woven into panels of mesh. The top edge of the net is held near the surface of the water by a line rigged with a series of plastic floats. The bottom edge of the net is attached to a weighted line. This float and weight setup causes the net to deploy like an underwater curtain 10 to 100 feet deep and 300 to 1,800 feet long, depending on water depth and the area fished. Fishermen place the net where they think schools of salmon will pass. The fish swim into the almost invisible net and become entangled by the gills and fins.

Nets are deployed from small boats, called gillnetters. If the gillnetter is set up to deploy the net from the front end, it is known as a bowpicker. If the net is deployed from the rear of the vessel, the boat is called a sternpicker.

Sternpicker.

Sternpicker.

Bowpicker.

Bowpicker.

Gillnetter

0'   5'   10'   *Sternpicker*

King Salmon

Pink Salmon

Coho Salmon

Herring

Chum Salmon

Sockeye Salmon

The fisherman clips a colorful numbered buoy to the end of the net, which is then paid out from the boat, usually perpendicular to the tide or current. The boat with net attached drifts with the tide, usually for an hour or more, depending on how many fish are hitting the net. A single drift may catch only a few fish or as many as a thousand fish.

Salmon gillnetters pick their fish one-by-one out of the net. Herring gillnetters remove their catch by vigorously shaking the net, usually with a power shaker or beater.

TONY LARA

A gillnet is hauled in with a good catch of sockeye salmon in Bristol Bay. Because salmon gillnets are deployed in the path of known schools of migrating salmon, there is little or no bycatch of non-target species.

Gillnetting is an efficient, relatively inexpensive way to fish. Many gillnet operations are husband and wife teams or whole families. The technique is attractive to both full-time fishermen and others, such as school teachers, who have time off from their regular jobs. Permits greatly vary in cost, depending on the area. For example, drift gillnetting from an open skiff in the Yukon and Kuskokwim rivers is much less expensive than gillnetting in less remote areas of Alaska.

Setnetting

Setnets are anchored gillnets used mostly in bays and near river mouths. One end of the net is anchored at or near the beach, and the other end is anchored 120 to 1,200 feet offshore perpendicular to the current.

Some setnetters deploy and retrieve their nets from small skiffs. Others simply wait for low tide and walk their nets out on the tidal flats and anchor them perpendicular to shore. At the next low tide, they walk back out and pluck the salmon out of their nets which are lying on the tidal flat. Then they load the catch into trucks or trailers pulled by all-terrain vehicles, and transport the fish to a processor.

In a few locations, like Cook Inlet, some setnets are connected to pulleys rigged like a clothesline. A truck, tractor, or power winch is used to drag the net onto the beach, where the fisherman picks out the salmon.

A setnet fisherman takes sockeyes
from his net in the Yakataga River.

Seining

The purse seine is another popular method of catching fish, primarily for catching schooling species, especially pink salmon and herring. Most seining occurs within a mile of shore. The boats usually are crewed by six people. The technique involves encircling a school of fish with a curtain-like net, then trapping the fish inside by closing the bottom of the net.

The first gasoline powered purse seine boat appeared in Alaska in 1910. From then until 1960, seiners and gillnetters competed with fish trap operations for a share of the Alaska salmon resource.

A purse seine crew brails chum salmon from a seine net at Hidden Falls in Southeast Alaska.

After the 1959 fish trap ban went into effect in 1960, people flocked to gillnetting and seining. By 1965, the purse seine fleet grew by 45 percent. Between the seine and gillnet fleets, in those five years about 6,000 new "mobile gear" (boat) fisherman entered the salmon fishery.

Purse seine fishermen may use sonar, a spotter plane or helicopter, or their own eyes to locate a school of migrating salmon or herring. When fish are found, the seine net is deployed and the fun begins.

The net, usually about 1,200 feet long, is deployed off the stern of the boat. Like a gillnet, the top of the seine net is attached to plastic floats and the bottom is weighted.

The skiff operator holds one end of the net as the seine boat makes a circle to surround a school of fish. When all the net is pulled off the seine boat and the circle is complete, the skiff operator passes the end of the net to crewmates on the seiner, closing the circle.

A set is usually held open for 30 minutes. Then a winch on the seiner pulls in a line which is strung

Purse seiner.

Purse seiner with net deployed.

Pink salmon from a seine net.

MANDY MERKLEIN

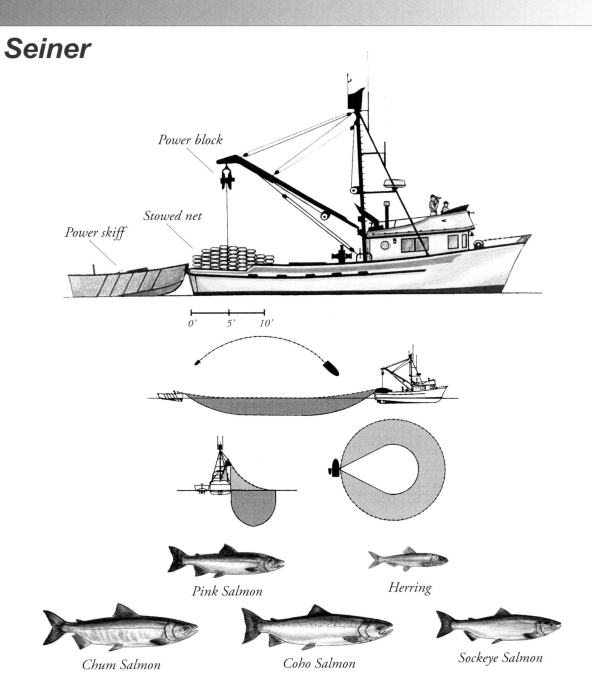

Seiner

Power block

Stowed net

Power skiff

0'   5'   10'

Pink Salmon

Herring

Chum Salmon

Coho Salmon

Sockeye Salmon

through metal rings attached to the bottom of the net. This closes the bottom of the net like a drawstring on a purse (thus the name "purse" seine), capturing the fish above.

Next the net is pulled aboard the seiner by a motorized winch, called a power block, which is mounted overhead on the seiner's boom. As the net is pulled in, the fish are massed in the last portion of the net, called the bunt or, more fondly, "money bag."

If it is a small haul, the bunt is hoisted aboard and the fish are dumped onto the deck or straight into the hold. If it is a big haul, the crewmen use a big dipnet, called a brailer, to dip the fish out of the seine net a few hundred pounds at a time.

Variations on purse seining include the hand seiner and beach seiner. A hand seiner is a smaller seine boat which uses the fisherman's muscle power instead of a motorized winch to purse and retrieve the net. Beach seiners make sets from the beach in very shallow water and do not use a purse line. Neither of these seine techniques is very common in Alaska and they account for a small fraction of the catch made by seiners.

Some seiners use smaller-mesh nets to fish for herring. Herring seiners usually do not take their fish on board the boat. Instead, a tender boat pulls alongside and lowers a suction pump hose into the seine net and sucks the fish into the tender's refrigerated hold.

CHARLIE ESS

A herring spotter plane taxis past seiners and tenders at Togiak.

Crewmen bring up the rings, closing the bottom of a seine net
that holds sockeye salmon at Cape Alitak, Kodiak Island.

Trolling

In Southeast Alaska salmon also are caught by trollers (not to be confused with "trawlers," described later in this chapter) on baited hooks or artificial lures. Clear water and the presence of feeding salmon migrating through the region make trolling feasible there.

Like gillnetters, trollers are often family businesses. They usually enjoy long seasons and the freedom to roam great distances up and down the Southeast Alaska coast. Trollers catch and handle fish individually, and they dress and ice or freeze them onboard. Because of this careful handling and because the feeding fish are in top condition, trollers are renowned for producing the highest quality salmon.

Trolling is confined to Southeast Alaska by state regulation, but compared with other Alaska salmon fisheries, this gear type is relatively unregulated. However, in recent years even trollers have been subjected to curtailed seasons through provisions of a salmon management treaty with Canada.

Two salmon trollers sit at anchor in one of the many bays on Baranof Island near Sitka.

A troller usually is a 25- to 50-foot boat rigged with one or two pairs of poles, each nearly as long as the boat itself. When not fishing, these poles are parked in a vertical position. When fishing starts, they are lowered to a 45-degree angle from the sides of the boat.

A set of hand- or hydraulic-powered reels, called gurdies, is mounted on the stern of the boat. The gurdies contain spools of long, fine, stainless steel wire. The line runs off the spool of the gurdy, over a block or pulley, and down into the water. While the wire is going out, the fisherman clips on the "spreads," which consist of a lure or baited hook, a leader, and a stainless steel snap. Lines extending from the tips of the poles, called taglines, are clipped to the wire after it is paid out, causing the wire to swing away from the boat, which prevents tangling. Sometimes a brass bell is attached to the top end of the tagline, which rings when a fish is hooked on a lure attached to the wire.

With the boat moving about two miles per hour, the wire is pulled under water to the proper depth by a lead ball weighing 15 to 60 pounds. After one set of wires is paid out, another set is deployed. Large foam plastic floats are clipped on in such a way that the second

Southeast Alaska troller.

A troll fisherman heads and guts
a freshly caught king salmon.

Troll-caught king salmon.

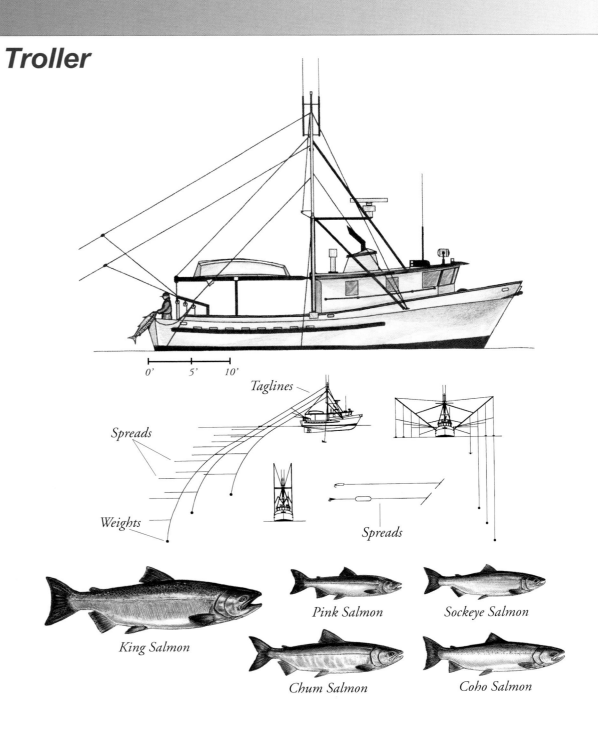

Troller

0'   5'   10'

Taglines

Spreads

Weights

Spreads

King Salmon

Pink Salmon

Sockeye Salmon

Chum Salmon

Coho Salmon

set of wires clears the first set and trails behind.

Using this system of poles, taglines, and floats, a power troller may set four lines in state waters and six lines in federal waters, towing 20 to 40 lures at once. Hand trollers are allowed to deploy two lines.

When a salmon hits one of the lures and becomes hooked, it yanks on the spread, which tugs the wire, which jiggles the tagline, which wiggles the pole or another small line called a tattletale, which runs to the bow of the boat.

Then a crew member engages the appropriate gurdy to bring in the wire and unclips each spread until the one with the fish appears. The fisherman stops the gurdy, pulls the fish up to the side of the boat with the leader, conks the fish on the head, gaffs it, and pulls it into the boat.

After the gear is reset, the fish is carefully gutted, washed, and laid in a bed of ice or slush (mixture of ice and water) in the fish hold, a prime fish ready for the seafood market, smokehouse, or fine restaurant.

A Southeast Alaska troll fisherman uses a hand-whittled gaff to land a 45-pound king salmon off Whale Island.

A long-in-the-tooth troller at Wrangell sports
a well-equipped business end.

Tenders

In many fisheries, the catcher boats do not run to town to deliver their catch directly to a processing plant. Instead they deliver to transport vessels, called tenders. Tenders, some of which can carry more than 100 tons of fish, may take deliveries from a dozen or more catcher boats before heading to the processing plant. In addition to delivering fish to the processor, the tender also brings fuel, drinking water, groceries, and other supplies to the fishermen. Some tenders offer hot meals and showers, cold drinks, and ice cream to fishing boat crew members.

Decades ago most tenders were sturdy wooden vessels, 60-100 feet long. They had large bunkers below deck where fish were hand-layered in bins and covered with crushed ice. But others were little more than flat scows or barges. Fish were dumped on deck without benefit

Two gillnetters enjoy a tender moment in Southeast Alaska.

Salmon tender.

Crabber converted
to herring tender.

Sorting salmon
aboard a tender.

Tenders

0'   10'   20'   30'

0'   10'   20'   30'

King Salmon

Herring

Pacific Cod

Coho Salmon

Halibut

Chum Salmon

Pink Salmon

Sockeye Salmon

of either shade or ice. These unchilled tenders, or "dry scows," are now a thing of the past.

While some wooden tenders still operate, most modern tenders are fiberglass or steel. Their holds are sealed, insulated, and refrigerated. Most common are seawater tank holds chilled by mechanical cooling, called refrigerated seawater systems (RSW), or by addition of crushed ice, called chilled seawater systems (CSW). Systems also are used that spray refrigerated seawater (spray brine) over the catch in the hold, or simply layer the fish in crushed ice.

GUY HOPPEN

Pink salmon on the deck of a purse seine boat are sucked into a ten-inch hose by a fish pump onboard a tender tied up to the seine boat.

Herring tenders take on their loads by extending hoses connected to large diesel-powered vacuum pumps into the fish holds of gillnet boats or into the pursed-up nets of seine boats. The fish are sucked directly into the tender's refrigerated tanks.

Vacuum pumps and hoses likewise are used to transfer salmon from seine vessel holds into tender holds.

Tenders usually take salmon from gillnet vessels in mesh bags, called brailers. The brailers are hoisted aboard the tender by articulating booms mounted on the tenders.

An electronic scale is attached between the end of the boom and the bag. Each bag is weighed as it is lifted aboard. A "tallyman" records on a "fish ticket" the weight and species composition of the catch. A copy of the fish ticket is given to the fishing vessel's skipper at conclusion of the delivery. The fish ticket is used for calculating payment to the fishermen, which usually occurs at the end of the season, and for reporting the catch to the Alaska Department of Fish and Game.

Many tenders are fishing vessels—usually longliners, trawlers, and crabbers—temporarily converted for tendering during salmon and herring seasons. When salmon or herring openings are finished, pumps and scales are removed and the boats return to their primary fishing operations.

Tenders may work a circuit of fisheries which can use up the majority of a year. Fishing

A seiner offloads pink salmon
to a tender at Kodiak.

vessels converted for tendering can spend more of the year tendering than fishing. A boat can start in April and May to tender the herring fishery in Southeast Alaska, Prince William Sound, Togiak, and Norton Sound; switch in June and July to tender Bristol Bay salmon; move to Southeast Alaska in August to tender pink salmon; head farther south to Puget Sound to tender a fall chum salmon opening; and finally refit for fishing operations in late fall and winter and head to the Gulf of Alaska or Bering Sea to harvest crab or bottomfish.

Halibut and Cod Fishing Gear

Longlining with baited hooks is the primary method used to harvest halibut, sablefish (blackcod), rockfish, Pacific cod (P-cod), and turbot. Blackcod and Pacific cod also are harvested with pots (traps), and some Pacific cod are taken with trawl gear. For an explanation of pot fishing, read the Crab Fishing Gear description that follows this section.

TONY LARA

A longline fisherman plays tag with a wave as he stretches to gaff a halibut near Kodiak.

A longline fisherman hauls in a Pacific cod in the Aleutian Islands. Pots also are used to harvest P-cod.

Longlining

A longline setup is built around a length of strong, thin rope called a groundline. The groundline comes in lengths of 1,800 feet, called a skate. Attached to each skate are up to two hundred stout leaders, called gangions (pronounced "GAN-yuns"), each with a baited hook. Spacing and size of the hooks depend on the target species. Blackcod gear uses smaller hooks spaced closer together than halibut gear. Hooks are baited with herring, octopus, or chunks of salmon. Several skates are tied together to make a string. The number of skates in a string varies with the type of fishing, depth, and size of the fishing vessel.

Fishermen haul up a halibut on the starboard side of a seiner that was temporarily converted for longlining in Prince William Sound during one of the now-defunct 24-hour halibut openers.

Each end of the string is held on the bottom by an anchor. A line runs from each anchor to the surface where brightly colored buoys are tied to mark the line for easy retrieval. To further increase visibility at sea, a bamboo or aluminum pole with a small flag, and a radar reflector or light, is added to the buoy rigs.

To set the gear, the fisherman first throws the flag into the sea off the stern of the boat. Then the buoy and the buoy line are paid out. When the buoy line is trailing straight behind the boat, the string anchor is dropped. Then the boat moves away in a straight line, which pulls the baited string into the water. As the end of the string is reached, the boat slows and the other anchor and buoy are set, leaving on the seafloor a string of baited hooks, anchored at each end.

Two kinds of bottom

A typical longline vessel.

Longline crew members deploy groundline over the chute.

Halibut are taken from a longline vessel's refrigerated hold at Cordova.

A 256-pound halibut ready for processing at a Cordova plant.

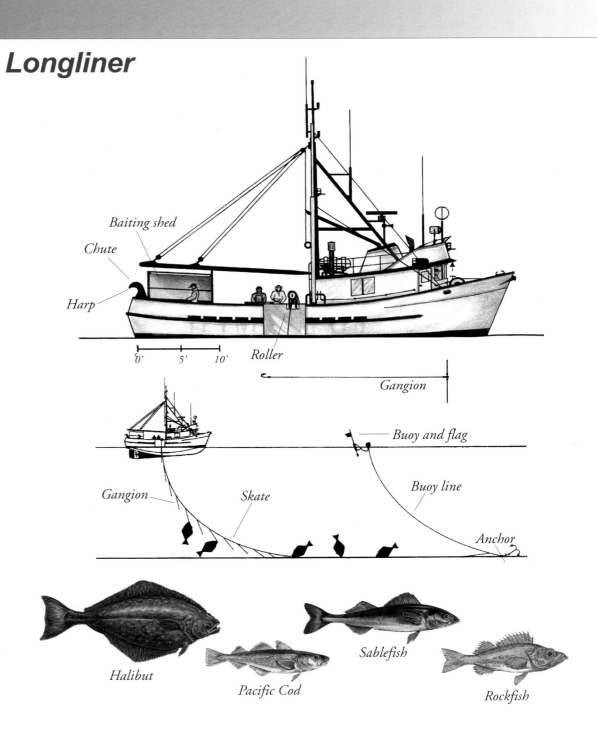

Longliner

Baiting shed

Chute

Harp

Roller

0'   5'   10'

Gangion

Gangion   Skate   Buoy and flag

Buoy line

Anchor

Halibut

Pacific Cod

Sablefish

Rockfish

(demersal) longline gear are used in Alaska fisheries: stuck and snap-on gear. Surface gear is used elsewhere, but not in Alaska fisheries.

Stuck gear has gangions permanently tied to the groundline, and is stored in tubs on deck ("tub gear") or coiled on square canvas sheets called skate bottoms, or diapers.

Snap-on gear incorporates stainless steel snaps to connect gangions to the groundline, which is stored on and retrieved by a hydraulically activated drum or reel. Hooks are baited in advance, then quickly snapped onto the groundline as it is paid out behind the boat.

Some of the larger longliners use automated systems, which consist of rack storage for the hooks on stuck gear and some type of automatic hook baiter. A good automated system requires fewer crew members and largely eliminates the tedious and time-consuming gear overhaul, the work of manually stripping old bait off the hooks, storing the hooks, and rebaiting them before they are set.

After all the gear is set, the vessel returns to the first string to start retrieving the gear. When it is time to haul back the gear, the captain steers the boat alongside one of the buoys

KURT BYERS

A longline vessel delivers its catch to a processing plant at Cordova. Halibut are cleaned, bled, and iced soon after they are taken aboard.

Gumby Suits and Satellites

The life-saving orange-red "Gumby" suit is made of closed-cell foam rubber.

Contestants swim to a life raft in a survival suit race at Kodiak.

Technology has provided new ways to escape otherwise fatal situations at sea. Next to radio (for calling help), the three inventions that have done the most to save fishermen from doom in the unforgiving northern ocean are the covered life raft, immersion suit, and the Emergency Position Indicating Radio Beacon (EPIRB).

The life raft has come a long way from its humble beginnings. Today, a well-appointed auto-inflating life raft has a waterproof canopy equipped with a locator light and rain catcher for drinking water, insulating floor for added warmth, boarding ramp, and a sea anchor and ballast pockets for stability in rough seas. They also include a survival equipment packet, which may contain flares, signal mirror, whistle, knife, fishing hooks and line, and other potentially life-saving goodies.

In Alaska's frigid waters, most people who fall overboard without protection against the cold water will die in less than an hour, even if they aren't injured or overcome by seas. Many unfortunate souls could have been rescued had killer cold not gotten there first. Even a well-equipped life raft may not be enough to prevent death by hypothermia. Enter the immersion suit.

The development in the 1970s of the immersion suit, also known as the "survival suit," put the odds back in the fisherman's favor. The bulky one-piece body suits are awkward to wear on deck, but have saved the lives of hundreds of fishermen who were forced to enter the water from a sinking, capsized, or burning vessel. A correctly fitted and worn immersion suit keeps the wearer afloat and prevents heat loss from all parts of the body.

Commercial fishing festivals sometimes feature races wherein contestants don immersion suits on a dock or shore, dive into the water, and swim to a life raft. The races are an effective way to highlight the use and lifesaving value of immersion suits and how to get into a life raft.

While the immersion suit keeps the fisherman alive while he is bobbing in the sea or sitting in a life raft awaiting rescue, the EPIRB makes rescue possible. EPIRBs are small radio transmitters about the size of a large flashlight. When activated, they transmit skyward an electronic beacon that retransmits off an orbiting satellite, down to a receiving station on earth. The signal alerts emergency responders that there is a problem, and guides U.S. Coast Guard search-and-rescue aircraft and vessels to the trouble spot.

and crew members pull the flagpole and buoy aboard. The buoy line is then fed through the powerful hydraulic hauler (or onto the drum in the case of snap gear), which pulls the anchor and groundline off the seafloor and onto the boat. The line usually is hauled in near the midpoint of the starboard (right) side of the boat.

As the line comes aboard, crew members gaff the fish and unhook or unsnap them from the groundline. Immediately they gut and wash the fish, and put them on ice or in slush in the boat's hold.

Crab Fishing Gear

Pot Fishing

Crab, Pacific cod, and blackcod are caught in cage-like fish traps, called pots. A pot consists of a welded steel frame covered with polyester or stainless steel mesh. A bait jar or sack full of chopped herring is hung inside to attract the crab or fish. If the fisherman is targeting crab, he also hangs a whole fish inside the pot to keep the crab busy until the fisherman has a chance to return and haul in the gear.

Lured by the bait, the fish or crabs find their way into the pot, but then either cannot find their way back out or can't get past the spring-loaded gate, called a "trigger," that allows passage in but blocks exit.

The size and style of pots vary quite a bit depending on the area to be fished and the species targeted. For example, Dungeness crab pots generally are cylindrical, about three feet in diameter and a foot high. A typical Dungeness pot weighs about 80 pounds and is covered with stainless steel

TONY LARA

Crew haul in a pot of snow crab from the Bering Sea.

Dungeness crabber at Kodiak.

Dungeness crab pot.

A fisherman measures a red king crab from Bristol Bay.

Crabber

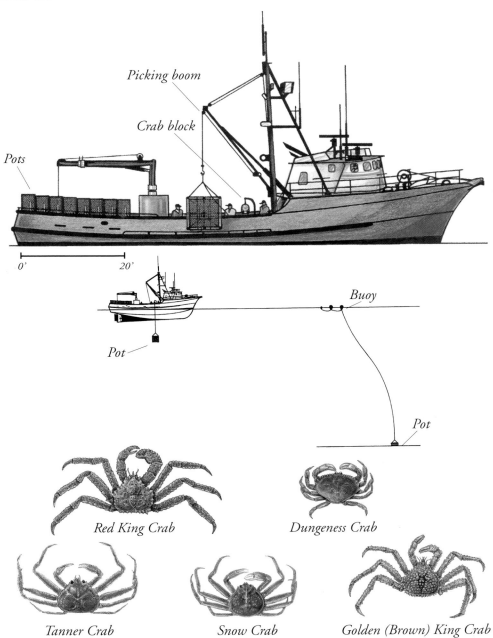

Picking boom

Crab block

Pots

0'   20'

Buoy

Pot

Pot

Red King Crab

Dungeness Crab

Tanner Crab

Snow Crab

Golden (Brown) King Crab

mesh. They are set individually and are marked with a buoy setup which consists of one or two large bullet-shaped corks. Dungeness crab are usually found close to shore and in protected bays, so look for Dungeness buoys in shallow water within three miles of shore. The typical Dungeness crab boat is 36-60 feet long which allows them to maneuver and operate safely in shallow water and small bays and channels. Dungeness seasons run from summer into late fall.

Snow crab, king crab, Tanner crab, and Pacific cod are fished with large, cube-shaped pots. The typical pot is 6.5 to 8 feet wide and 3 feet high. An empty pot weighs around 700 pounds and must be brought aboard the crabber with a powerful hydraulic winch and crane. Each pot is marked with two or three large colored buoys attached to a line connected to the pot on the seafloor.

An exception to this is golden king crab (also known as brown king crab), which are fished along the Aleutian Island chain in water as deep as 3,600 feet. To increase efficiency on a region of the seafloor which is exceptionally rough and irregular, fishermen use a single line with about 40—but sometimes up to 75—pots attached at 600-foot intervals. The idea is much like longlining. Instead of hooks, steel pots are attached to a 1.5 inch groundline, and no anchors are needed. The pots are unhooked from the groundline when they are hoisted aboard the vessel.

TONY LARA

Just another day at the office as Bering Sea crab fishermen haul in a pot full of snow crab.

There are several other styles and variations of the longline/pot rigs. Hair crabbers, for example, use small pots spaced about 60 feet apart. The big cubical pots and cone-shaped Tanner pots are the ones most commonly rigged this way.

In most crab fishing, pots are set out individually, each marked with a buoy. Depending on the type of fishing, the average soak time is about 24 hours. After the desired soak, the captain steers the boat so the crew can snag the buoys with a grappling hook. The buoys are pulled aboard and the pot line is fed into the hydraulic-powered block. A crab block uses the same principle as a longline hauler, except it handles larger line and reels the line in much faster.

When the pot breaks the surface, the block is stopped and a crew member hooks the pot to a cable connected to a crane-like mechanism (called the picking boom) which lifts the

Beating the Odds

TONY LARA

Crab fishing in the Bering Sea is tough and often dangerous work.

For decades commercial fishing in Alaska has been the most dangerous occupation in the United States. A fatality rate of 140 people per 100,000 workers per year was 20 times the national average.

In the late 1970s, the high casualty rate caused liability insurance rates to soar and forced the fishing industry to get serious about finding ways to reduce risk. In 1983 the Marine Advisory Program of the University of Alaska developed a commercial fishing safety outreach program. An outgrowth of that effort was formation of the Alaska Marine Safety Education Association (AMSEA) in 1985. AMSEA provides marine safety training and safety instructor certification.

Also in 1985, the North Pacific Fishing Vessel Owners' Association (NPFVOA) started a vessel safety program for their members.

In addition, the U.S. Coast Guard instituted voluntary safety guidelines, but most fishermen ignored them. It would take a personal disaster and the crusade of a bereaved and angry woman to give the safety movement the boost it needed.

In 1985 a college student named Peter Barry drowned when a seiner he was working on sank off Kodiak. His mother, Peggy Barry, went to Congress on a personal mission, determined to improve safety aboard fishing vessels.

In 1988, Congress—thanks in part to Peggy Barry—passed the Commercial Fishing Industry Vessel Safety Act. The law mandated that vessel owners get survival suits, life rafts, and institute safety training. AMSEA, NPFVOA, the Coast Guard, and other entities were ready to provide help.

The combined efforts are paying off. While the number of vessels lost in the Alaska fishery remained pretty steady through the 1990s, the fatality rate dropped by more than half.

Some segments of the industry—notably the offshore crabbing and trawl fisheries—are still dangerous places to work. But safety awareness, knowledge, and skills are now more widespread and prospects are good for further improvement.

pot over the rail and onto a hydraulic rack (pot launcher). From here, the pot door is opened and the crab are dumped into plastic bins (crab totes) or onto a sorting table. Crew members sort the crab, returning undersize crab and females to the sea and putting the "keepers" into the boat's hold. Meanwhile, the pot is rebaited and put back in the water, or the lines and buoys are tied up and stored in the pot which is then stacked on deck.

Once aboard the fishing vessel, the crab are kept alive in flooded holds, which are large tanks, until just minutes before they are cooked by the processor. Seawater is pumped through the tanks to keep the water clean and oxygenated.

TONY LARA

A couple of fishermen show off a big red king crab taken from the Bering Sea.

The majority of king, snow, and Tanner crab fishing in Alaska takes place during the winter far offshore in the Bering Sea. This is the time of year when crabs are most able to withstand being caught and tossed back into the sea if they do not meet harvest size and sex requirements. To safely operate in frigid, often stormy conditions, the average Bering Sea crabber today is 100- to 180-feet long, with some large catcher-processors in the 250-foot range.

Newer, larger boats, coupled with increased safety awareness, have made crab fishing much safer than it was in the late 1970s and early 1980s. A new quota share management system diminishes the dangers related to a competitive fishery, and openings now can be delayed to avoid fishing during bad storms. Still, crab fishing remains one of the most dangerous occupations in the United States due to frequently adverse weather and the heavy, cumbersome gear that stressed crew must manipulate on pitching, slippery decks.

Pots are one of the "cleanest" (little bycatch mortality) fishing devices because the fish or shellfish are brought aboard the boat alive, and undersized crabs and other bycatch can be returned to the sea with little or no harm.

However, lost pots can continue to catch fish and crabs, a problem dubbed "ghost fishing." Regulations call for pots to be constructed with sections made of rapidly biodegradable or corrodible materials so that lost pots do not continue to capture animals indefinitely.

A crewman, known as a "stack monkey," ties down snow crab pots aboard the F/V *Reliance.* Four boats sank in four days and one fisherman died during this 1994 winter storm on the Bering Sea.

Groundfish Fishing Gear

Trawling

Most cod, pollock, sole, and other bottomfish—also known as white fish or groundfish—are harvested by trawlers. Trawlers are large (60 to more than 300 feet), powerful vessels which tow funnel-shaped nets along the bottom (bottom trawlers) or up in the water column (midwater trawlers). These two trawl fisheries are generally considered to be separate fisheries. A third kind of trawl fishing is used to harvest shrimp. (See "Shrimp Fishing Gear" in this chapter.)

Until the early 1970s, only a small number of American trawlers fished Alaskan waters, delivering to shore-based processing plants. But in 1976, sweeping changes in U.S. law forced foreign fishing vessels out of the 200-mile U.S. fishery zone (see Fisheries Management chapter). By the late 1970s, the only foreign fishing vessels allowed into the zone were

FRED HIRSCHMANN

A trawler heads from Frederick Sound to Wrangell Narrows in the Inside Passage of Southeast Alaska.

A typical trawler.

A small trawler underway at Kodiak.

Trawl net drum and door.

Bottom Trawler

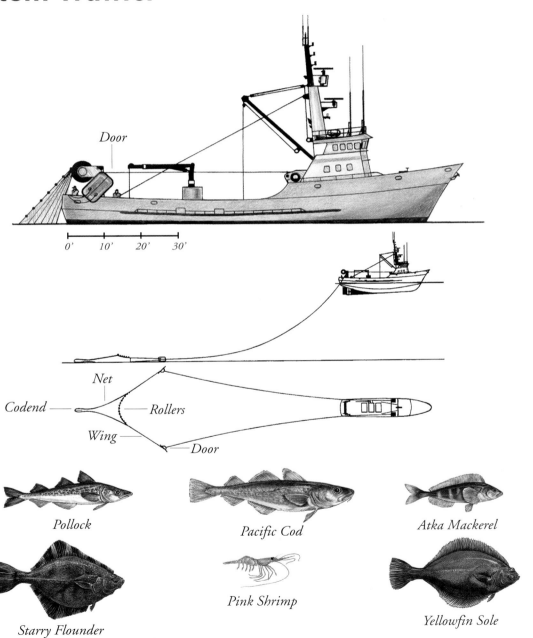

Door

0'   10'   20'   30'

Net

Codend — Rollers

Wing — Door

Pollock

Pacific Cod

Atka Mackerel

Starry Flounder

Pink Shrimp

Yellowfin Sole

processors, which purchased fish caught by American boats. Eventually the U.S. processing industry built enough capacity to handle the catch, and foreign processors were displaced.

Now the harvest is processed by a mix of shore plants, floating processors, and factory trawlers—also known as catcher-processors—which are equipped to both catch and process groundfish. As of 2000, about 20 American-flagged catcher-processor trawlers fished Alaska waters. Most of them are about 140 to 200 feet long and employ up to 50 people onboard at a time.

Trawl nets are rigged to the fishing vessel with heavy steel cables, and may span from 100 feet to 400 feet at their opening, depending on the size and power of the vessel. Crew members pay out the trawl net and cable over the stern until the net is at the proper depth. The vessel's skipper monitors sophisticated electronic devices which show him where the fish are in front of and under the boat and in relation to position of the net. Sensing equipment also indicates the vessel's speed and the amount of fish in the net.

Fish trapped in the net soon succumb to exhaustion and drift to the narrow rear end of the net, called the codend. When enough fish are packed into the codend, the skipper orders

KURT BYERS

Trawlers tied up at Kodiak await the next trip. Notice the Steller sea lion hauled out on the stern ramp. Although the big animals can be a nuisance to fishermen, federal law prohibits harassment of this endangered species.

Midwater trawler.

A codend full of pollock rises from the Bering Sea.

A crewman washes pollock into the holding tank aboard a Bering Sea trawler.

Midwater Trawler

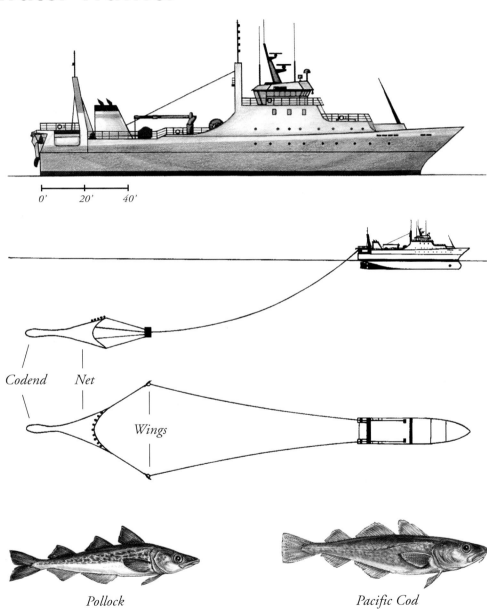

0'   20'   40'

Codend   *Net*

Wings

Pollock

Pacific Cod

the crew to engage winches which retrieve the cable and slide the net up the stern ramp.

Bigger boats use nets that can catch over 100 tons of fish in a single tow. On some vessels the codend is opened on deck and the fish are shoveled into holds below deck for delivery to shoreside processing plants. On others the codend is towed to a nearby mothership or other factory trawler where the fish are winched aboard the second ship and processed on the spot.

NATALIE FOBES

A midwater trawl net opens to discharge thousands of pollock from the Bering Sea.

Bottom trawl gear targets flatfishes and other species that hug the bottom. Foot ropes, sometimes made of chain, attached to the net travel along the seafloor. In some cases, hard rubber rollers are used to protect the net and help it pass over rocks and other obstacles.

"Chafing gear," made of heavy rope matting, protects the codend from abrasion as it passes over the seafloor. Steel panels, called doors, are attached at either side of the net opening. As the boat pulls the net through the water, the force of the water on the doors pushes them outward, spreading the net open. The turbulence caused by the doors on the seafloor stirs up sediment clouds as the net travels through the water, herding fish into the net. A big bottom trawl net spans about 200 feet in width with an opening about 12 feet high.

Midwater trawl nets are much larger. They do not run on the seafloor so they exert less drag on the vessel, allowing the boat to tow a larger net with the

Goodbye Bycatch?

A federal fisheries observer estimates the volume of pollock in the codend of a trawl net.

In most Alaska fisheries, technology and personal knowledge allow fishermen to efficiently find and harvest from large, concentrated single-species populations—resulting in what's called "clean" fisheries. But no matter how carefully fishermen fish or how advanced harvest technology becomes, there will always be some bycatch taken from the sea—fish, invertebrates, birds, and mammals that are not the intended target. The trick is to minimize bycatch while maintaining the economic viability of the fishing industry.

Bycatch in the groundfish fishery is an ongoing challenge. Each year fishermen sweep more than 3 billion pounds of pollock from the Gulf of Alaska and Bering Sea. In 1996, a typical pollock harvest year, 96 percent of the catch was pollock. That's the good news. The bad news is, the remaining 4 percent bycatch still amounted to about 132 million pounds of fish, a lot of wasted protein.

To make matters worse, fishermen used to be required by law to throw away certain species of bycatch (called "prohibited species"), dead and wasted. Finally, in 1999, after years of heated debate a regulation was adopted which requires trawlers to keep and process a certain amount of non-target species.

Longliners sometimes hook seabirds when they flock around lines to snatch an easy meal off hooks before the baited hooks sink out of reach into the sea. Field trials by scientists working with foreign and U.S. longline fishermen have succeeded in developing new techniques to scare birds away from longlines and to get the baited lines to sink faster. Efforts continue to find still better techniques to avoid catching seabirds.

In the crab fishery, thousands of lost crab pots are believed to "ghost fish," continuing to trap and kill crabs and fish on the seafloor. Crab pots are required by law to be constructed with biodegradable panels. But they may not break down fast enough to release alive all bycatch.

These are a few of the bycatch problems concerned parties are grappling with. While excellent progress has been made through better laws and fishing techniques, intensive work continues by fishermen, scientists, technicians, conservationists, and managers to reach their elusive goal—to avoid catching non-target and undersized species in the first place.

same power. A 1,800 horsepower midwater trawler might tow a net with a rectangular opening 240 feet wide and 180 feet high. A net with an opening that size can be up to 1,000 feet long, including the codend.

Some trawlers are propelled by as much as 4,000 horsepower, although their nets are not necessarily proportionally larger than those of medium-power vessels.

Midwater nets are equipped with doors but don't funnel fish into the net like bottom trawl nets. They have huge openings and large-mesh "wings" at the outside corners of the opening. The wings herd free-swimming fish into the net's gaping maw.

New high-density polyethylene ropes, which are extremely light, thin, and strong, are used to make the nets because the ropes reduce the drag of wings and nets in the water, allowing a vessel to sweep a larger volume of water.

Some huge midwater nets include a detachable codend that allows the vessel to tow its catch to a shore plant, mothership, or factory trawler without taking the fish on board.

Jigging

One of the oldest forms of fishing in the world, jigging, became one of the newest in Alaska when the state opened a quota-regulated nearshore jig fishery for Pacific cod in 1997. The season opens January 1. Most fishing occurs in the spring and summer until quotas are filled.

Jigs are used on small boats from 20 to 58 feet long to catch Pacific cod, rockfish, and other groundfish. The technique involves lowering one or more artificial lures on a line and

TONY LARA

rapidly jerking the line up and down, either manually or mechanically, to impart motion to the lures and induce fish to strike.

Traditionally jigging was done by hand. But now semi-automated jigging machines have revolutionized small-boat bottom fishing. Reels are powered by electric or hydraulic motors controlled by a small computer. The computer finds the proper depth, makes the reel jig the lures up and down the right distance and speed, alerts the fisherman when there are fish on the line, and on the fisherman's command instructs the reel to bring them up. A small boat with one or two crew members is allowed to operate up to five machines at once.

Jigging machines are particularly effective for catching spawning and feeding aggregations of Pacific cod and rockfish. Permits cost only $50. A computerized jigging rig costs anywhere from about $1,000 to $5,000.

A Kodiak fisherman using an electric jigging machine pulls in some rockfishes.

High-Tech Fishing

The wheelhouse of a modern trawler is equipped with an array of electronic equipment that improves fishing operations. In the face of onrushing technology that constantly increases fishing efficiency, the North Pacific Fishery Management Council, fishermen, and other stakeholders work together to find equitable ways to ensure the long-term sustainability of Alaska's offshore fisheries.

Modern trawlers with their sophisticated sonars, global positioning systems, computers, weather fax machines, powerful diesel engines, and highly trained and motivated crews are the top predators on the sea. To stay profitable amid ever-shortening fishing seasons and increased competition, fishermen must make the most out of their boats.

Bob Desautel is one of those fishermen. He runs a medium-size, 165-foot trawler, 38 feet wide. It can hold about 500 tons of fish.

"We fish mainly in the Bering Sea, sometimes in the Gulf of Alaska. We have five crew members and one federal fisheries observer aboard. We primarily fish for pollock, but we also fish for cod, Atka mackerel, yellowfin sole, and crabs. The rest of the time we find other revenue sources, such as salmon and herring packing and other things to keep the boat busy.

"Our wheelhouse is outfitted substantially with electronics," explains Desautel. "We've got three radars, three plotter systems, and three sonars. We've got two down-sounders and a third-wire system that goes down with the net to detect fish that enter the net and tells us how much quantity is in the net." And because the law requires that trawlers keep all the fish they catch, Desautel's nets feature specially woven panels along the top of the net to allow small, undesired fish to escape.

Desautel and fishermen like him drop about a quarter million dollars worth of gear in the water when they are towing. With millions of dollars on the line every season, fishermen cannot afford to have anything go wrong when their vessels are hundreds of miles from land.

"The way the fisheries are now, you need to have backups and redundancies. You miss a trip and there goes any kind of profit. We have backups for our backups. I've got two autopilots and three steering systems, spare main engines, spare generators, spare cranes, spare this, spare that. I've even got spare sonars and computers sitting in my stateroom. It's pretty sophisticated stuff."

Fishing isn't what it used to be.

Shrimp Fishing Gear

Beam Trawling

Similar in concept to offshore groundfish trawls but much smaller in scale, beam trawls (with many local variations) are used to harvest shrimp in Southeast Alaska. The technique involves towing a small net, which is held open underwater by a 30-foot wood or aluminum beam.

A beam trawl is dropped to make another tow for pink shrimp. This beam is made of hemlock, which sinks better than spruce. Each tote holds about 150 lbs of shrimp.

Beam trawls were developed by the Japanese in the 1800s, and have been used in Alaska since about 1916. They are especially well-suited for confined and rocky irregular fishing grounds common in Southeast Alaska. Because they can be rigged on small one-person vessels, such as sternpicker gillnet boats, beam trawls continue to be popular and effective for shrimp fishing.

Historically, otter trawls with heavy steel spreader doors, rather than beams, took large quantities of pink shrimp in Alaskan waters. But this fishery virtually disappeared with the crash of the pink shrimp stocks in the 1970s and 1980s.

Pot Fishing

Pot fishing for shrimp is smaller in scale but similar in technique to pot fishing for crabs. Shrimp pot fishing occurs mostly in Southeast Alaska and primarily targets spot shrimp, *Pandalus platyceros,* one of several species of large shrimp generically known as "prawns." The fishery yields less than 1 million pounds per year. Most is sold locally and in other states, but some is marketed in Japan.

Boats usually are seiners, gillnetters, and trollers rigged for pot fishing. The pots are deployed on longlines with about 20 box-shaped pots attached. Limits are set on the number and size of pots each fisherman is allowed to deploy.

Habitat Is Key to Industry's Future

BRENDA KONAR

Fishermen increasingly understand the need to preserve habitat that is essential to the fish and shellfish they depend upon for their livelihoods.

Alaska's marine and coastal zone habitats are among the most pristine in the world, but even here deliberate as well as unintentional acts have left their mark.

The most dramatic case of habitat degradation was the 1989 Exxon Valdez oil spill, which poisoned the water and fouled beaches from eastern Prince William Sound to lower Cook Inlet, Kodiak Island, and as far west as Chignik on the Alaska Peninsula.

A less cataclysmic, but more persistent form of habitat degradation is the effect of timber harvest practices on fish rearing habitat, particularly in Southeast Alaska. Timber harvest and associated road construction may choke streams with silt, deprive stream banks of essential cover and food for young fish, and change water temperatures. Log storage fouls bays and estuaries with decomposing woody debris.

Under pressure from environmental groups and federal legislation, the timber industry has greatly reduced its impact on fish habitat, and a changing timber market and protective laws have decreased the annual cut. Major pulp mills in Ketchikan and Sitka, which for decades polluted the water, closed in the 1990s. The prospect is that a scaled-back timber industry which uses appropriate harvest methods can be a good neighbor to Alaska's fisheries.

Virtually invisible, but far-reaching in its effect on fisheries and their supporting habitat, is the damage caused by fishing gear, particularly the bottom trawl. The nets themselves—large, heavy bags of synthetic mesh—scour the seafloor, rake the sand, dislodge boulders, and snap off corals, anemones, kelps, and any other living substrate they contact. The nets are spread by heavy steel "doors" that scrape the bottom to a depth of several inches.

Bottom trawls not only damage or diminish the productivity of some habitat, they also take large quantities of bycatch—fish, shellfish, and other creatures that are prohibited or have no economic value to the fishermen and are returned to the sea, usually dead. Bottom trawling has been banned in some areas, and by 2000 a movement had gained momentum to forbid the gear in any fishery where an alternative gear type exists for taking the target species.

Pound Fishing Gear

Herring spawn by laying millions of eggs on kelp fronds and other underwater brush in subtidal and intertidal waters. Fishermen in Prince William Sound and Southeast Alaska take advantage of this behavior to harvest herring eggs without killing herring. In a fascinating harvesting technique, fishermen build floating corral-like enclosures with bottoms, called pounds, which they fill with wild *Macrocystis* kelp they take from selected beds in Southeast Alaska. The kelp fronds, which often are hauled in from locations many miles away, are tied to webbing inside the anchored pound and are suspended in the pound from lines strung across the top.

Purse seiners then catch nearby schools of herring, tow them to the pounds, and release the fish into the pound.

If the fishermen are lucky and their timing is right, the captive herring will lay their eggs on the kelp fronds. After that, the herring are released unharmed to the sea, and the roe-laden kelp leaves are removed from the pound, trimmed, packed in salt, and exported to Japan, fetching premium prices.

DAVE GORDON, ADFG

A herring roe-on-kelp pound is ready for harvest in Hoonah Sound.

An ADFG technician inspects *Macrocystis* kelp leaves attached to a line in a pound. The kelp is covered with herring roe.

Scallop Fishing Gear

Weathervane scallops, also known as giant Pacific scallops, are taken by dredges towed along the sea bottom by fishing vessels typically about 100 feet long. In most regulatory districts scallopers are limited to a maximum of two dredges, one deployed from each side of the vessel, each with a maximum width of 15 feet. The dredge is made of a heavy welded steel frame attached to a bag of steel rings and nylon mesh. The frame glides over the bottom on steel runners called "shoes." A sweep chain on the forward end disturbs the bivalves, which use rapid water expulsion to propel themselves off the bottom, and into the path of the dredge. A complete dredge weighs up to a ton.

JEFFREY BARNHART, ADFG SCALLOP OBSERVER PROGRAM

A crewman dumps bags of scallops onto the deck from two dredges.

Dredges are set and retrieved with heavy steel cables deployed by large hydraulic winches. Product processing and freezing is usually done aboard the boat. The scallops are hand-shucked, and only the large adductor muscles (which close the shell) are retained for sale.

Dive Harvest Gear

Urchins, sea cucumbers, and geoducks are three potentially valuable invertebrate species that are harvested by divers using a technique known as hookah diving. "Hookah" is a transliterated word for "water pipe," derived from the Middle Eastern Urdu language. In the fishery context it is just a nickname, not a precise descriptive term. Abalone also used to be harvested by divers, but the abalone fishery is closed due to excessive predation by sea otters.

Hookah fishing gear includes a small tender boat from which a heavily weighted diver clad in a dry suit descends to the harvest area, usually 30 to 60 feet underwater. Compressed air is pumped to the diver through a hose attached to the diver's equipment. Divers pick the urchins, abalone, geoducks, and sea cucumbers by hand from their niches and put them in a net bag. No mechanical harvest tools are allowed other than a hand-held rake for urchins, and no more than two permit holders are allowed per boat during dive harvest openings.

The technique for geoduck harvest is not quite as straightforward. Geoduck clams live

Geoduck diver with "stinger."

An adult and juvenile sea urchin.

A batch of sea cucumbers is transferred to another tote.

Hookah Dive Gear

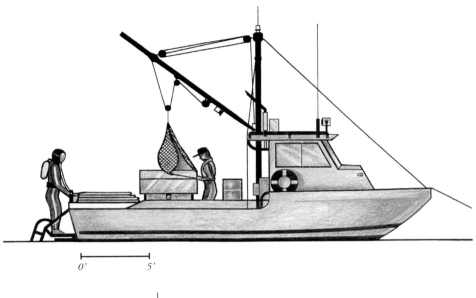

0'   5'

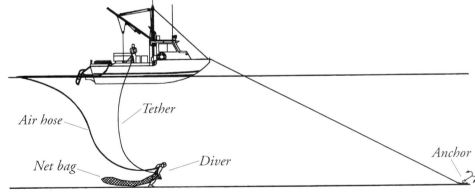

Air hose

Tether

Net bag   Diver

Anchor

Sea Cucumber

Urchin

Geoduck

burrowed about two to three feet in sandy seafloors. These big bivalves, averaging two to four pounds with the biggest ones up to 20 pounds in weight, extend their siphons (necks) up through the sandy mud to suck in water from which they filter out food particles. The siphons create telltale "dimples" in the seafloor, which divers look for as they move along the bottom.

To extract the clam, a diver uses a hose that blows a high-pressure jet of water through a hand-held wand, called a "stinger." The diver uses the stinger to blow away bottom sediment, exposing the geoduck's neck. Taking care not to bruise the siphon, the diver grasps the geoduck's neck and pulls the giant clam out of the sandy seafloor.

CHARLIE ESS

A fisherman hauls up a net bag full of geoducks as his partner tends the diver's air hose.

Divers sometimes use scuba gear to harvest urchins and sea cucumbers. Scuba often is preferred when the harvest area has a lot of kelp or other underwater hazards that could entangle hookah lines.

Some dive harvest fishermen carry both scuba and hookah gear aboard their boats and use whatever is best for the harvest location they find.

But scuba diving should be done with a partner and scuba gear is more costly to maintain than hookah, all of which adds greatly to the cost of doing business. Due to cost of equipment and maintenance, and the longer time hookah gear allows a diver to stay underwater, hookah is the gear of choice in the dive harvest business.

A geoduck diver displays
his formerly buried treasure.

A tote full of sea urchins is moved on a Southeast Alaska dock.

Clamming

There is a small commercial razor clam harvest along the west shore of Cook Inlet. Only about 30 people harvest the clams. Most clam diggers travel each year to Alaska from Mexico and the western United States, and spend the summer in temporary camps near the beach.

The low-tech gear includes a modified spade, net bag, hip boots, tender rafts, plastic buckets, and sometimes a wood shoulder yoke to haul filled buckets. The harvesters walk along the beach looking for dimples in the beach sand which indicate the presence of clams. With one or two quick strokes they dig up the clams and put them into their net bags.

When the bags are full the clams are put into plastic buckets in the tender raft. An airplane lands on the beach several times a day, collects the harvest—which amounts to about 1,000 pounds per plane load—and delivers the clams to a processing plant. About 370,000 pounds of razor clams are harvested each summer, sold mostly in the western United States. Diggers also take a small quantity of hardshell clams in some Alaska locations.

CHARLIE ESS

Clammers dig razor clams near Polly Creek
on the west shore of Cook Inlet.

Common Alaska Fishing Vessels

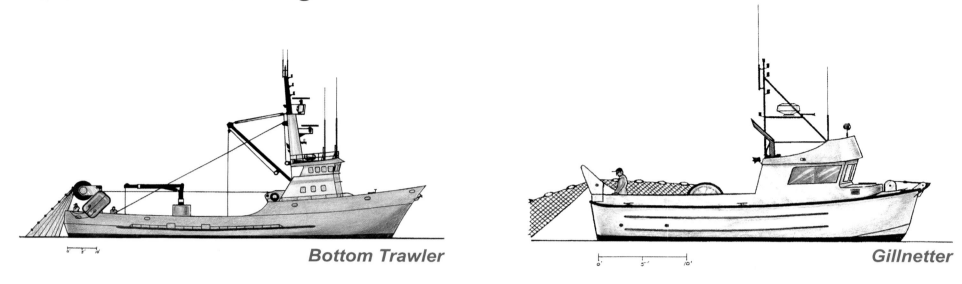

Bottom Trawler

Gillnetter

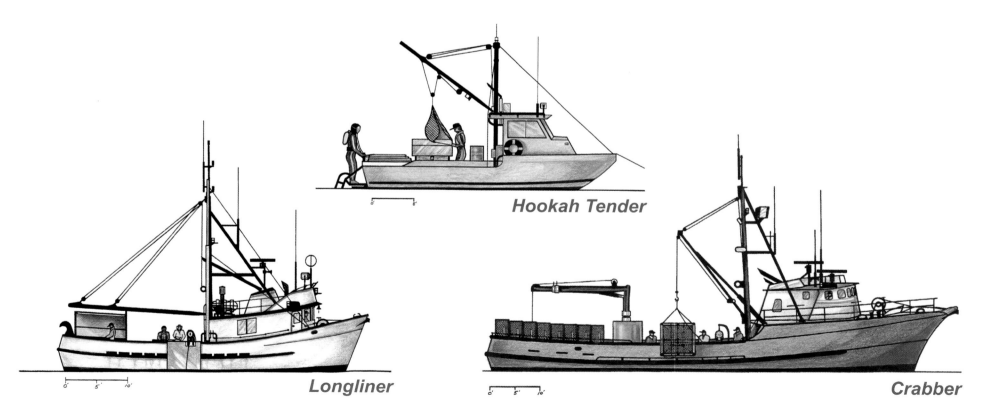

Hookah Tender

Longliner

Crabber

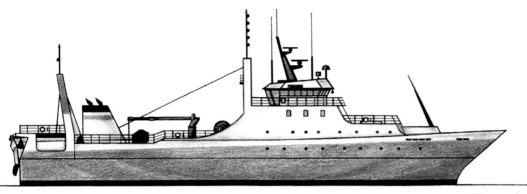

Trawler

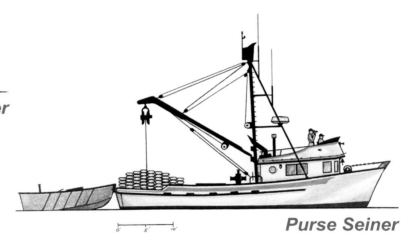

Purse Seiner

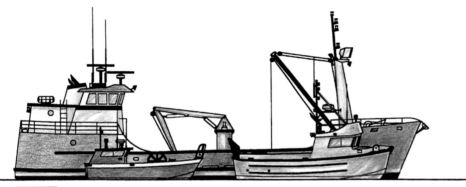

Tender

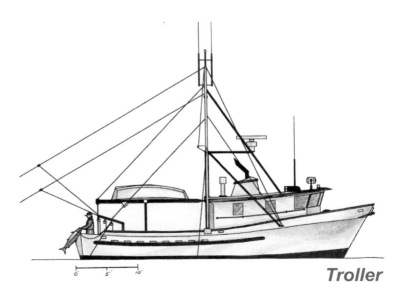

Troller

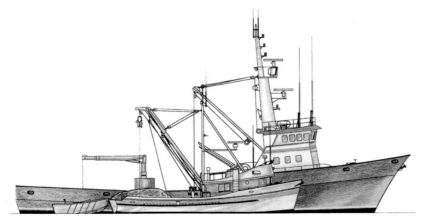

SE tender

Chum salmon appear in full nuptial dress as they return to a Juneau hatchery. The color patterns catch the attention of potential mates and rivals.

FINFISH AND SHELLFISH CULTURE

While the harvest of wild fish and shellfish from the world's oceans appears to have reached a plateau, artificial propagation—commonly known as aquaculture and mariculture—is steadily increasing throughout the world. Mariculture of shellfish is rapidly expanding in Alaska, and Alaska shellfish mariculturists have been pioneers in cold water mariculture.

The terms aquaculture and mariculture sometimes are used interchangeably. Strictly speaking, mariculture refers to raising and harvesting fish, shellfish, or sea vegetables in salt water, usually in some kind of pen constructed in coastal waters. The fish and shellfish are never released to the open sea. This approach also is called "farming." When carried out in freshwater ponds or lakes, this propagation technique is generally called aquaculture.

When animals are propagated from native stocks in hatcheries and released as juveniles to mature in the "pasture" of the open ocean and return to be harvested in the common property fishery, the technique is called ocean ranching.

Shellfish Culture

Alaska's developing shellfish farming industry traces its roots to the 1930s when Pacific oysters were planted on beaches in the Ketchikan area. Although various attempts at bivalve farming were made in the 1950s and 1960s, it was not until the late 1970s that modern suspended culture farming took root in Alaska.

MARILYN HOLMES

A mariculturist tends a lantern net that contains oysters at a suspended culture oyster farm near Craig, Alaska. Suspended culture involves the use of lantern nets—cylindrical cages of mesh stretched over steel hoops. The cages are suspended in the water from floats.

Bivalve production farms are located in Southeast Alaska, Prince William Sound, Resurrection Bay near Seward, and Kachemak Bay near Homer. In addition, some on-bottom farming of littleneck clams occurs in Southeast Alaska, and the Aleutians will host mariculture farms in the 2000s.

The next frontier for Alaska shellfish farmers is on-bottom culture of clams, scallops, geoducks, cockles, sea urchins, and several edible and medicinal kelps. On-bottom culture usually involves leasing a section of beach or intertidal area from the state, partitioning the area with protective net fences and covers (to prevent predators from eating the cultured animals that are placed there), and letting the animals grow to harvestable size over several years.

By 2000, Alaska's shellfish farming industry was selling nearly half a million dollars worth of product per year. And the potential for growth is enormous as the state carefully regulates development.

Johnny Appleseed Revisited

Oyster brood stock at the
Qutekcak Hatchery.

A flupsy unit in Halibut Cove
near Homer.

Some shellfish—oysters in particular—grow well in Alaska's cold, nutrient-rich waters. But oysters do not reproduce well in Alaska, if at all. Because of that, oyster farming in Alaska could not advance until a reliable source of oyster seed—called spat—could be found.

The solution came with money from the Exxon Valdez oil spill settlement, some of which was used by the state in the mid-1990s to build the Qutekcak Shellfish Hatchery in Seward. The goal was to establish an in-state capability to produce spat for shellfish restoration projects, sale to oyster farmers, and aquaculture of native clams and scallops. And because importation of spat for native bivalve species was prohibited by law, Alaska needed an in-state facility that could produce spat for native species such as littleneck clams. The facility is operated by the Qutekcak Native Tribe. As of 2003, the Qutekcak Hatchery was the only source of bivalve spat for Alaska mariculturists.

To produce spat, technicians induce brood oysters to spawn by regulating photoperiod (amount of time the oysters are daily exposed to light), water temperature, and diet. Together these factors trigger a sensory response which causes the oysters to produce larvae. The larvae are then put into large tanks where they mature into spat.

When the spat reaches a certain size, it is transferred to other tanks where it is placed on whole or fragments of oyster shells, depending on the process. The spat continues to grow on the shells or shell fragments until it is mature enough to be sold to oyster farmers.

Viable spat is valuable stuff. To minimize mortality and get maximum productivity after it reaches the field, several groups of oyster farmers have collaborated with the University of Alaska aquaculture specialist to develop specially designed bivalve nursery units to enhance growth and survival of spat. The technology is called a floating upweller system, or "flupsy."

A flupsy mariculture system uses a pump to draw nutrient-rich seawater into cubical or cylindrical containers partially submerged in water, in which the spat is held. The constant water flow increases the amount of nutrients available for the spat which accelerates growth, and removes bivalve wastes from the containers. The result is lower production costs and increased survival of young bivalves for transfer to open-water racks where they continue to grow to marketable size.

Ocean Ranching

Commercial finfish mariculture is prohibited in Alaska by an act of the state legislature to protect wild salmon from disease and genetic impacts, and Alaska commercial salmon fishermen from additional market competition. While finfish mariculture is illegal, another kind of artificial propagation, called ocean ranching, is used extensively to bolster Alaska's salmon stocks. The intent of ocean ranching is to provide for the majority of the returning fish to be caught by sport and commercial fishermen in the "common property" fishery.

Similar to mariculture, ocean ranching involves incubating and hatching native stock salmon eggs in hatcheries. But instead of keeping the fish in pens until adulthood, the juvenile salmon are released to the wild in specially selected streams and estuaries. From there, they swim out into the sea where, depending on the species, they spend one or more years feeding and maturing with wild salmon in this huge ocean "pasture."

When the salmon mature and are ready to spawn, they instinctively migrate back to their place of birth. As the ocean-ranched fish approach the hatchery, river, or estuary where they were released, they are harvested by commercial and sport fishermen.

Catch quotas are set to ensure that enough of these returning salmon make it back to the hatchery to provide eggs and milt to produce another year's crop of ocean ranched salmon. Fish surplus to spawning needs are taken in "cost recovery" fisheries that produce revenues to help cover hatchery expenses.

In contrast with salmon ranching, salmon farming—which is legal elsewhere in the world but not in Alaska—involves incubating, hatching, and raising salmon to marketable size in pens without releasing the fish to the ocean. When the farmed fish are large enough to market they are taken from the pens, processed, and sold to seafood wholesalers and distributors.

CHARLIE ESS

A hatchery worker ratchets in a net that funnels salmon into the hatchery processing facility. Inside, eggs and milt are taken to spawn another year's output of salmon.

Depending on the species and water temperature, after hatching, salmon alevin feed themselves from their yolk sacs from about 14 to 90 days.

Salmon Hatcheries

Alaska has a successful ocean ranching program carried out through a network of 31 private nonprofit hatcheries (PNPs), two sport-directed state hatcheries, and three federal hatcheries, strategically located around the state. Most are in Southeast and Southcentral Alaska.

The network is made up of small nonprofit corporations (sometimes called "mom-and-pops") and larger regional aquaculture associations, all of which are licensed to hatch and rear salmon for release into the sea.

Alaska has a distinguished record of success in the field of nonprofit ocean ranching, which helped revitalize the state's salmon fishery when it was near collapse in the early 1970s. Alaska's salmon hatchery system is second only to Japan's as the world's most productive, and is generally viewed favorably by its chief beneficiaries, commercial fishermen. However, in recent years some observers have criticized the program for producing too many fish that end up competing in the marketplace with wild-stock salmon.

Several small-scale private nonprofit hatcheries are operating, most of them in Southeast Alaska, providing jobs and modestly contributing to the overall salmon landings of the region. But the big successes have been the regional PNP associations, usually made up of commercial fishermen who assess themselves either 2 or 3 percent of the ex-vessel price to help fund the capital costs and operations of the hatcheries. The PNP hatcheries have succeeded in raising all five species of Pacific salmon. In fact, they have been so successful at producing salmon that the State of Alaska decided in the late 1980s to turn over its hatcheries to PNP operators.

CHARLIE ESS

Hatchery workers at a private nonprofit salmon hatchery in Prince William Sound take eggs from pink salmon.

Banned in Alaska

A technician feeds fish meal to fingerling salmon at a Prince William Sound hatchery. After growing in the hatchery through the winter, the small salmon are released to the ocean or ocean tributaries in the spring when plankton, the juvenile salmon's food source, is plentiful near shore.

Commercial mariculture of finfishes and, to a lesser extent, ocean ranching of salmon are controversial topics in Alaska. During the 1980s, fish farmers in other parts of the nation experienced some success raising salmon in pens. The pristine protected coastal bays and estuaries of Alaska looked like an ideal place to expand pen-rearing of salmon for commercial use. Recognizing the potential for salmon farming in Alaska, people who thought its time had come tried to change Alaska laws to permit salmon farming in the state.

The effort failed in the face of stiff opposition from Alaska commercial fishermen and others who feared market competition, water pollution, spread of disease, and weakening of the wild salmon gene pools by escaped pen-reared salmon. In 1990 the Alaska legislature passed a law that prohibits commercial finfish farming in Alaska.

Meanwhile salmon farming expanded dramatically in neighboring British Columbia. In recent years fishermen and scientists in Southeast Alaska have encountered increasing numbers of Atlantic salmon that escaped from Canadian net pens, and B.C. scientists have reported that, at least in that province, the escapee fish are successfully reproducing in the wild.

Alaska law does allow farmers to raise oysters, mussels, and clams in Southeast Alaska, Prince William Sound, and Kachemak Bay. Experiments are under way to raise scallops, geoducks, and red algae seaweed (nori).

State mariculture regulations have undergone substantial modification since shellfish farming began in the late 1970s. Farmers can now apply for long-term leases of state tidelands, which vastly improves their opportunities for financing. Research and development is actively being pursued with significant progress being made in production of shellfish seed, nursery culture to accelerate growth and shorten the growing cycle, and development of new farming techniques. Shellfish production is growing each year and all indications are that it will continue to expand.

The hatcheries now release approximately 1.5 billion juvenile salmon into the ocean each year. In some years about 80 percent of the pink salmon landed in Prince William Sound originate in hatcheries operated by the Prince William Sound Aquaculture Corporation. PNP-reared salmon also add millions of dollars worth of fish to landings in Southeast Alaska and Cook Inlet.

The regional associations do more than just hatch and rear salmon. Much of their effort is directed at enhancing natural runs through the use of incubation boxes placed in streams, and through stream rehabilitation, such as clearing logjams and building fish passageways around barrier falls.

Too Many Salmon?

There is no doubt that Alaska's salmon hatcheries helped save Alaska's salmon fishery in the 1970s and have since contributed enormously to the world's greatest salmon fishery, which generally remains robust as other fisheries around the world suffer serious declines. But there are concerns.

Chum salmon churn up the raceway upon their return to a Juneau hatchery.

For example, some scientists wonder if Alaska's and other nations' salmon hatcheries have been too successful, putting so many salmon into the sea that they may be disrupting the dynamics of the ecosystem.

Another, much less speculative, problem has occurred in recent years as farmed salmon have taken over a larger portion of the world salmon market and ocean ranched and wild salmon populations have remained plentiful. The difficulty is that huge numbers of ocean ranched and wild Alaska salmon combine with farmed salmon from around the world to create an oversupply. When the market is glutted, prices plunge, processed fish clog storage facilities, and fisherman cannot make enough money to cover expenses—to the point where some opt to stay in port even when huge numbers of salmon are available (see sidebar).

The Great Humpy Dump

Pink salmon near the end of their life cycle at the Kitoi Hatchery in Kitoi Bay at Afognak Island.

As farmed salmon have gained an increasingly large share of the world salmon market, the Alaska seafood industry has faced tough sailing in trying to market the millions of salmon it harvests each year.

In the most extreme example, in 1991 Alaska experienced a record return of pink salmon—128 million fish. This included record or near-record returns to all three of Alaska's primary pink salmon fisheries of Kodiak, Prince William Sound, and Southeast Alaska. Prices paid for pink salmon, about 12 or 13 cents per pound, ex-vessel value, were so low that hundreds of fishermen refused to fish and boycotted the salmon openings.

In Prince William Sound, without harvest by fishermen, millions of pink salmon simultaneously converged on the hatcheries where they were spawned. Even if all fishermen had gone fishing, processing capacity was not large enough to handle all of their catch.

In an emergency response to protect water quality around the hatcheries, the hatcheries hired fishermen to harvest some of those fish and take them—2.8 million of them—out to sea to dump. Otherwise, close to 4 million dead pink salmon would have fouled the water near hatcheries, potentially creating anaerobic "cesspool" conditions in violation of the federal Clean Water Act and other state and federal environmental quality standards.

Not all of those salmon were wasted, however, in what is remembered as the "humpy dump" or "pump and dump" year. Some 32.6 million pink salmon were harvested as usual in Prince William Sound and made it to market, albeit at bargain-basement prices. In addition, in a scramble to find ways to use some of the salmon, the State of Alaska donated about 1.3 million canned pink salmon to the Alaska Food Bank and to the Soviet Union, which was in the throes of a food shortage. Another 164,000 pink salmon were shipped to Kodiak to be used in research on production of salmon surimi.

Since then, the seafood industry and food scientists have increased efforts to find different ways to process and market pink salmon so that the sale of pink salmon is not so heavily dependent on moving canned product.

Dense schools of thousands of herring
provide submarine banquets for seabirds,
marine mammals, salmon, and other fish.

FISHERY RESOURCES

Salmon long reigned as king in the Alaska fishing industry. But in recent decades the overall fishery has become quite diversified, with crab, shrimp, groundfish, and other species making up important segments of this $925 million* industry.

The offshore groundfish industry is the largest single component of Alaska's fishery. However, for most Alaskans the salmon fishery is more important because salmon are harvested by small- to medium-scale fishermen, many of whom live in Alaska and buy equipment and services in Alaska. From 1995 through 1999, the ex-vessel value (money paid to fishermen) of Alaska's salmon harvest averaged about $345 million. From 1995 to 2001, the yearly catch averaged about 170 million salmon, including an all-time record harvest of nearly 218 million salmon in 1995, a record almost matched in 1999 with a harvest of close to 217 million fish. However, as market competition forces salmon prices down, total value has decreased. And in the early 2000s, there is evidence of a decline in salmon stocks, probably mostly due to a change in ocean water temperature.

Groundfish are primarily taken by large companies based in Washington state. This fishery yielded about $510 million in ex-vessel value each year from 1994 to 1999. The huge offshore fishery is based on Alaska (walleye) pollock, Pacific cod, yellowfin sole, Atka mackerel, blackcod (sablefish), and rockfish. Pollock accounts for the largest portion of this fishery.

Most of the groundfish are taken outside Alaska's waters, from fishing grounds in the Gulf of Alaska and the Bering Sea, and never touches Alaska shores.

*2001 ex-vessel value, including bycatch, from Alaska Department of Fish and Game, Commercial Fisheries Division.

For the first 20 years following U.S. takeover of the groundfish fishery, relatively little income from that enterprise reached Alaska. That changed significantly in the 1990s when Community Development Quotas (CDQs) went into effect, channeling a percentage of the groundfish harvest to people living in tiny communities in the Bering Sea coastal region nearest the fishing grounds. See the Fisheries Management chapter for details on CDQs.

Halibut and herring are important money-makers for the inshore fleet, followed by other finfish species like Pacific cod and blackcod.

Shellfish are major revenue producers. Once-dominant king crab is now overshadowed by less valuable but more abundant snow (opilio) crab in the Bering Sea. However, in 1999 snow crab stocks plummeted and that fishery was scaled back. Dungeness crab continue to produce values in the millions of dollars each year.

Shrimp stocks in the Gulf of Alaska and Bering Sea are greatly diminished from levels of the 1970s. Beam trawlers still harvest a small quantity of mostly pink shrimp around Wrangell and Petersburg, and a modest pot fishery exists for the larger coonstriped and spot shrimp—better known as prawns—primarily in Southeast Alaska. Fisheries are developing for other invertebrates such as sea urchins, sea cucumbers, and geoducks.

The industry generally divides commercially harvested species into three categories: groundfish, shellfish, and non-groundfish finfishes, here referred to as "finfish."

Following is a short inventory of Alaska's fisheries resources. Most of the natural history information in this chapter comes from the Alaska Department of Fish and Game *Wildlife Notebook Series.* See the Sources and Useful References section in this book for more good sources of information on Alaska fish and invertebrates.

Finfish

Salmon

Five species of Pacific salmon inhabit Alaska's waters. Salmon comprise the most important Alaska commercial fishery because, unlike most of the offshore fishery, salmon are harvested and processed in Alaska. Salmon also support huge sport and subsistence fisheries.

The largest and least abundant salmon is the king (chinook) salmon, *Oncorhynchus tshawytscha.* Adult fish usually weigh between 15 and 50 pounds, but a 126-pounder was taken in a fish trap near Petersburg in 1949.

Juvenile chinooks in fresh water feed on plankton, and later on, eat insects. In the ocean, they eat herring, sand lance, squid, crustaceans, and other animals. Salmon grow rapidly in the ocean and often double their weight during a single summer. They return to natal streams to spawn after spending from two to seven years in the ocean.

Kings are found along the entire Alaska coast from Southeast Alaska to Norton Sound.

A sockeye (red) salmon returns to spawn in a river at Katmai National Park near King Salmon.

Annual commercial landings in the 1990s averaged 564,000 fish.

Sockeye (red) salmon, *Oncorhynchus nerka*, contribute the greatest value to the commercial fishery, based largely on a high per-pound price and huge runs in Bristol Bay. Reds average about six pounds each. They spawn in streams or lakes where they spend one to three years before migrating to the ocean. After an additional one to four years in the ocean they return to natal streams to spawn and die.

Sockeye salmon eat zooplankton, benthic amphipods, and insects, and in the ocean add fish and squid to their diet. In the mid-1990s, Alaska fishermen landed from 40 to 65 million sockeyes per season. Since then the harvest has declined a bit.

Coho (silver) salmon, *Oncorhynchus kisutch,* are caught in nearly every salmon fishing district of the state. They average eight pounds apiece, and are important to the industry

MARION OWEN

A coho (silver) salmon makes its way up the Buskin River on Kodiak Island.

Chum salmon in spawning coloration.

Pink salmon.

Chinook salmon (left) and two sockeye salmon.

A spawning coho salmon. The prominent jaw hook is called a kype.

King (Chinook) Salmon
Oncorhynchus tshawytscha

Chum (Dog) Salmon
Oncorhynchus keta

Coho (Silver) Salmon
Oncorhynchus kisutch

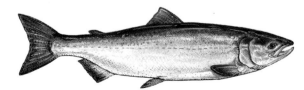

Sockeye (Red) Salmon
Oncorhynchus nerka

Pink (Humpback) Salmon
Oncorhynchus gorbuscha

even though annual landings only averaged around five million fish a year in the 1990s.

Most young silvers spend two winters rearing in freshwater before migrating to sea as smolts. Time at sea varies, but most silvers stay in the ocean for 16 months before returning to spawn as adults.

Chum (dog or keta) salmon, *Oncorhynchus keta,* are usually a little larger than coho but bring a lower price due to the pale color of their meat. They are the major species in the northern areas of Kotzebue and Norton Sound, but contribute to the fishery in nearly every district. An average of about 17 million chums were caught each year in the 1990s.

Unlike king, silver, and red salmon, chums do not stay long in freshwater after hatching. Instead they migrate to sea within a month or two of emerging from the gravel in the spring.

Chum fry feed on small insects in streams and estuaries before gathering into schools in salt water, where their diet usually consists of zooplankton. By fall they move into the Bering Sea and Gulf of Alaska, and return to natal streams to spawn and die after 3 to 6 years in the ocean. Adult chums average from 7 to 18 pounds, with females usually smaller than males.

Pink (humpback or humpy) salmon, *Oncorhynchus gorbuscha,* are the smallest but most numerous salmon, averaging about 4 pounds. They are caught by the tens of millions in Southeast Alaska, Prince William Sound, and around Kodiak Island. Along with chums they command the lowest prices, but are a mainstay of the canning industry. Through the 1990s an average of 103 million humpies were landed each year in Alaska.

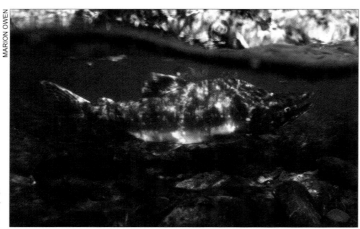

MARION OWEN

A male pink salmon nears the end of the line in Kitoi Bay on Afognak Island.

Most pink salmon spawn in streams within a few miles of the coast, in stream mouths, or in the intertidal zone. Following entry into salt water, juvenile pink salmon are easy targets for predators as they swim in dense schools near the surface along shorelines, feeding on plankton, larval fishes, and insects. By fall juvenile pink salmon are 4 to 6 inches long and move into the Gulf of Alaska and the Aleutian Islands area. At two years of age they migrate back to their natal sites to spawn.

All Pacific salmon die after they spawn. The carcasses are a food source for other animals and contribute essential nitrogen to lake or riverine ecosystems. The nitrogen helps keep the water bodies biologically productive to nourish the next generation of salmon.

Pink salmon in the final phase of their life cycle
mill around at a hatchery in Southeast Alaska.

Herring and Herring Roe

The Pacific herring, *Clupea pallasii*, is part of a worldwide family of fishes that makes a major contribution to the global food supply. Alaska's herring harvest averages around 50,000 tons a year. Herring fisheries occur in districts from Southeast Alaska to Norton Sound. The Togiak district of northern Bristol Bay produces nearly half of Alaska's entire catch, but large harvests also occur at Sitka, in Prince William Sound, and around the Kodiak archipelago. Smaller harvests occur near Ketchikan, Dutch Harbor, and along the Bering Sea coast.

Pacific herring inhabit the entire northern rim of the Pacific Ocean, including the Bering Sea. Over most of that range individual herring average 6-8 inches long and weigh only a few ounces. Bering Sea herring tend to be large and can reach 12-15 inches in length and about a pound in weight.

Herring return each spring to specific shores to spawn on kelp in the shallows. Eggs hatch in a few weeks and the larvae drift with the current, to grow and mature in the open ocean. In earlier decades, trawlers scooped up thousands of tons of feeding herring at sea. Now fisheries are conducted only close to shore.

ALASKA SEA GRANT

Herring milt appears as light colored areas along this shoreline near Sitka.

Sand lance.

U.S. COAST GUARD

Herring roe on kelp.

DAVE GORDON, ADFG

Herring eggs.

STEPHEN JEWETT

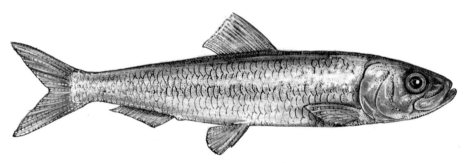

Pacific Herring
Clupea pallasii

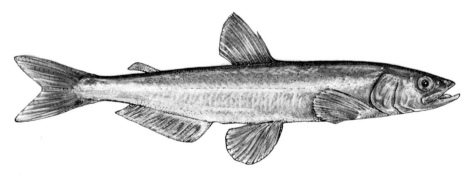

Eulachon (Hooligan)
Thaleichthys pacificus

Some Alaska herring fisheries produce fillets for human food or frozen bait for catching halibut and crab. But most are roe fisheries; openings are scheduled to harvest the fish just before they are ready to spawn. The skeins of eggs removed from the female carcasses are processed into a salted product that sells as a traditional Japanese New Year holiday delicacy known as kazunoko.

The humble herring's contribution to Alaska's fisheries goes far beyond simply providing roe to the Japanese market. Herring is an important "forage fish" whose oil-rich flesh supports other sea life. Herring is a key intermediate link in the food chain. They accumulate energy stored by zooplankton (which in turn accumulate energy produced by phytoplankton, the primary producers) and pass it along to the upper trophic level predators like salmon and halibut, marine mammals, and seabirds.

A seiner passes through a cloud of herring milt. Spawning occurs in the spring when male herring release their milt among millions of eggs which stick to seaweed, brush, and other things in shallow water. The eggs hatch in about two weeks, depending on water temperature.

JOHN HYDE, ADFG

Other forage fishes, like capelin and sand lance, perform a similar role. One that is particularly appreciated by many Alaskans is the eulachon, or "hooligan." These six-inch, smelt-like anadromous fish surge up certain coastal rivers in great schools during the early spring, where they are harvested by scores of Native subsistence fishermen and recreational dipnetters.

Southeast Alaska Natives used to trap large quantities of eulachon and boil them in big caldrons to render the oil, or "grease." Hooligan grease was an important foodstuff prior to the advent of packaged foods. Now most hooligan fishermen either fry the little fish fresh or smoke them to eat like jerky.

Groundfish

While the per-pound prices of salmon, crab, and shrimp are greater, the sheer mass of groundfish—also referred to as white fish or bottomfish—is so large that total groundfish values from Alaska have eclipsed those of the other fisheries. Groundfish landings for the Gulf of Alaska and Bering Sea total about 4.4 *billion* pounds of fish! Of that total, pollock alone accounts for more than half.

Forage Fishes and the Food Chain

Forage fishes comprise a broad category of small, silvery schooling fish including herring, capelin, eulachon, and sand lance, as well as juvenile cod and walleye pollock. These fish reside smack in the center of the marine food chain. They are eaten by a host of predators including salmon, pollock, halibut, arrowtooth flounder, marine mammals, seabirds, and people. These small fishes feed on plankton and are a key part of the process of converting phytoplankton to the flesh of other fish, seabirds, and marine mammals.

In addition to playing a key role in keeping the top of the food chain healthy, several of the species are commercially harvested, yielding millions of dollars in income. Eulachon, also known as "hooligan," are dip-netted by subsistence and recreational fishermen as dense schools of the fish migrate up streams to spawn.

We know a lot about commercially harvested forage species, primarily herring and pollock, especially the adult life stage. But we know less about their early life stages (eggs, larvae, and juveniles). And for species not commercially fished, such as capelin, eulachon, and sand lance, we know very little.

One of the things we do know, however, is that abundance of forage fishes has profound effects on the rest of the marine ecosystem.

Abundance is controlled in large part by ocean temperatures and other natural factors. Ocean conditions shifted in the late 1970s, which included a rise in ocean temperature. These natural changes, collectively known as a "regime shift," seem to have benefited pollock, flatfish, and herring, and perhaps caused a decrease in capelin, sand lance, crab, and shrimp in the Gulf of Alaska in the 1980s. Scientists suspect that this decrease in forage fishes contributed heavily to a subsequent decline in the 1980s and 1990s of marine mammals and seabirds, whose diets depend heavily on highly nutritious forage fishes.

While regime shifts appear to be the main reason behind fluctuations in forage fish abundance, commercial harvest and abundance of predators such as whales also may be factors. With the advent of sophisticated computer modeling capabilities, scientists are only now beginning to consider the entire ecosystem, including both human-induced and natural changes, in their daunting quest to understand fluctuations in fish populations.

These two images from a University of Washington video show (top) a coho salmon and (bottom) a puffin attacking a school of herring in Prince William Sound. Cohos have been observed to spread out and charge through schools of forage fishes, then gobble up fish separated from the school.

"Groundfish" is an imprecise term. Most are demersal (live on or near the ocean floor), and some are pelagic and live higher in the water. But they are still called groundfish because of the way they are caught, how they are used, and how they are managed.

Groundfish include Pacific cod, pollock, blackcod (sablefish), turbot, Atka mackerel, Pacific ocean perch, lingcod, and several species of flounder, sole, and rockfishes.

Most groundfish are taken by trawlers. Some vessels are equipped to process their catch. Other boats deliver their catch to processing plants on land. The great majority of groundfish is processed into frozen white fish blocks, used to make frozen seafood products such as fish sticks and the boneless fillets commonly found in supermarket frozen food departments. Fast food restaurant chains, economy seafood restaurants, and fish-and-chips restaurants also are major users of Alaska groundfish.

Pollock

The single greatest biomass of commercially exploited fish in Alaska waters is Alaska (walleye) pollock, *Theragra chalcogramma*, sometimes spelled "pollack." A small, slender relative of cod, pollock aggregate in enormous schools that can be hundreds of yards wide, hundreds of feet deep, and dozens of miles long. Large trawlers catch them tens of tons at a time.

Processing vessels convert them into frozen blocks of white flesh for use in fast food restaurants, or process them into a flavorless fish paste, known as *surimi*, for later conversion into manufactured food products like artificial crab legs. Pollock also are taken for their roe which is processed into *mintaiko* and *tarako*, both delicacies in Japan.

Pollock reach maturity at age four, about the same time they recruit (grow large enough to be economically desirable) into the fishery, when they are about 15 inches long. In the spring they migrate from the outer continental shelf to shallower water to spawn. Pollock reproduce and grow rapidly, and can adapt to a variety of habitats. They feed on plankton and small fish, and are eaten by birds, marine mammals, and other pollock. Juvenile pollock can be up to 80 percent of the food eaten by adult pollock in fall and winter.

In the late 1990s, an average of 2.7 *billion* pounds of pollock were harvested each year, worth more than $260 million, ex-vessel value.

Pacific Cod

Target of the first commercial fishery in Alaska, Pacific cod, *Gadus macrocephalus* (also known as true or gray cod, or P-cod), is still recognized as an important resource. Aleut hunters provided cod to Russian fur traders before the start of the salmon fishery. In the early decades of the twentieth century, dory schooners took cod for salting along the Alaska Peninsula and Aleutian Islands.

For decades the fishery was dominated by foreign countries. But a new domestic cod fishery

A juvenile walleye pollock lingers
in Southeast Alaska waters.

ART SUTCH

BILL BARSS, OREGON DEPARTMENT OF FISH AND WILDLIFE

Blackcod (sablefish) range from the Bering Sea and Aleutian Islands south to Baja California, and in the western Pacific Ocean from Japan north to the Commander Islands. These blackcod were photographed off the Oregon coast.

has developed on the grounds once worked by the schooners. Most cod are taken by trawlers and longliners, but a small-boat fishery recently has developed in nearshore waters, taking cod with jigs and pots.

Cod mature at about 25 inches in length when they are about six years old, and spawn in late winter. They eat clams, worms, crabs, shrimp, and small fish.

Cod landings in the late 1990s averaged over 570 million pounds per year. Average value to fishermen was around 15 cents per pound for trawl-caught fish and up to 60 cents a pound for pot- and hook-caught fish.

Several other Alaska species have "cod" in their names, but are not members of the Gadidae or cod family. Blackcod is a deepwater fish unrelated to cod. Lingcod, a bottom dweller in the greenling family, is another. Rock cod is the common term used for several species of rockfishes, spiny-back denizens of rocky bottoms.

Blackcod

Blackcod, or sablefish, *Anoplopoma fimbria,* is a schooling bottomfish with smooth, bluish-black skin. It is renowned for its rich white flesh. Steamed smoked blackcod is one of the finest meats from the sea, and is a premium product in Japan and other countries.

Blackcod mature after five years at about two feet in length, spawn during the winter, and may live more than 60 years. They inhabit the continental slope at depths as great as 900 feet. Blackcod feed on invertebrates, squid, and small fish.

Fishermen take blackcod with longlines, trawl gear, and pots along the entire Gulf of Alaska coast and in the Bering Sea. Landings range between 4 and 9 million pounds a year. Fishermen receive from less than one dollar to more than four dollars per pound for this highly sought species.

Blackcod.

Pacific cod.

Walleye pollock.

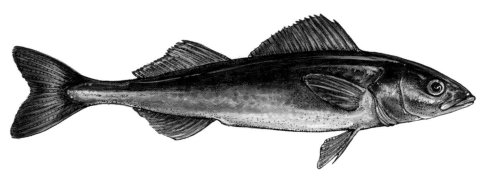

Blackcod (Sablefish)
Anoplopoma fimbria

Walleye Pollock
Theragra chalcogramma

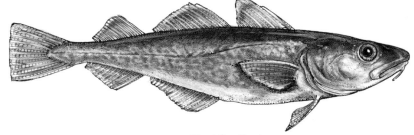

Pacific Cod
Gadus macrocephalus

Atka Mackerel

Not a mackerel, but a member of the greenling family, Atka mackerel, *Pleurogrammus monopterygius,* is a small (about 12 inches) schooling fish found in the North Pacific and Bering Sea, especially along the Aleutian Islands. They live on the edge of the continental shelf and migrate to shallow coastal waters to spawn during the summer. Nests of fertilized eggs are guarded by the male. Atka mackerel eat plankton and are eaten by marine mammals, birds, and fish. They enter the fishery at age two, first spawn at age three or four, and may live 14 years.

Atka mackerel are taken by trawlers. Yearly catches of this groundfish have ranged from 33 to 176 million pounds with an ex-vessel value of around 15 cents per pound. They are popular in Asia, but not common in American cuisine.

ROBERT R. LAUTH, ALASKA FISHERIES SCIENCE CENTER

A bumblebee-colored male Atka mackerel guards a nest of eggs in 75 feet of water near Seguam Island in the Aleutians.

Rockfishes

Thirty-four species of rockfishes occur in Alaska waters. Several species contribute to commercial fisheries. Depending on the species, they range in size from less than one to over three feet in length. Most are one to two feet long and four to ten pounds at maturity. Almost all rockfish species bear live young. Rockfishes are known to have long life spans, but do not reproduce rapidly. For example, shortraker rockfish, *Sebastes borealis,* and rougheye rockfish, *Sebastes aleutianus,* are believed to live 120 to 240 years.

The directed rockfish fishery in Southeast Alaska targets yelloweye rockfish, *Sebastes ruberrimus.* Yelloweyes are sometimes called "red snapper," although they are unrelated to the true snappers that inhabit seas farther south.

Another commercially important rockfish is a species known as Pacific ocean perch, *Sebastes alutus* (which are not perch). Pacific ocean perch, like most rockfishes, are slow-growing and long-lived, with a life span up to 90 years. Pacific ocean perch school in waters of 300-1,200 feet in the Bering Sea and Gulf of Alaska.

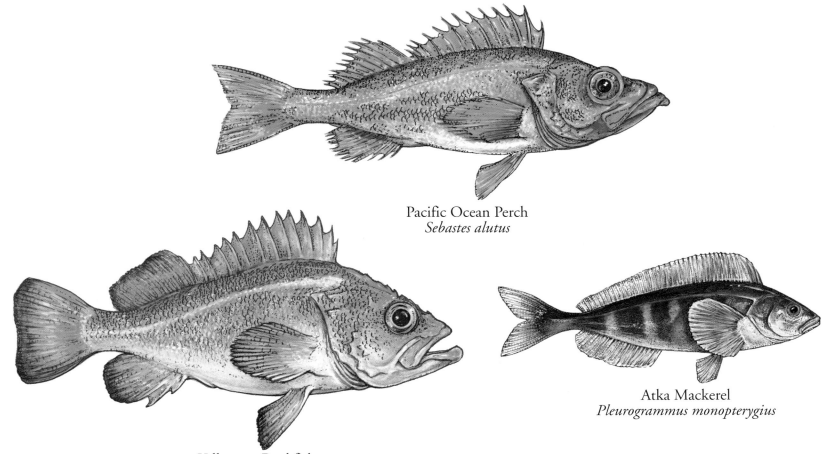

Atka mackerel.

Yelloweye rockfish.

Black rockfish.

Pacific Ocean Perch
Sebastes alutus

Yelloweye Rockfish
Sebastes ruberrimus

Atka Mackerel
Pleurogrammus monopterygius

Pacific ocean perch were heavily exploited by foreign fleets in the 1960s when catches peaked at about 330 million pounds. But in recent years *combined* rockfish landings have totaled about 55 million pounds. Ex-vessel prices have averaged around 19 cents per pound.

In Western Alaska, the black rockfish, *Sebastes melanops*, also known as black bass, is the most important commercially harvested rockfish species, with about 750,000 pounds harvested each year in the late 1990s.

Rockfish harvest is carefully controlled because the long-lived species do not reproduce rapidly and little is known about the size of the overall population.

Flatfishes

Several species of flatfishes, commonly referred to as soles and flounders, are taken in Alaska's groundfish fisheries. All are bottom dwellers, living mainly on sandy or gravel bottoms in the Bering Sea and along stretches of the Gulf of Alaska coast.

After halibut, the most important of Alaska's flatfishes is the yellowfin sole, *Pleuronectes asper*. Yellowfin sole are small fish, averaging about 13 inches long, but exist in huge numbers in shallow Bering and Chukchi sea waters to 300 feet deep. Harvested by trawlers, annual landings in the late 1990s averaged some 236 million pounds.

Landings of other flatfishes include rock sole, *Pleuronectes bilineatus*; flathead sole, *Hippoglossoides elassodon*; arrowtooth flounder, *Atheresthes stomias*; Dover sole, *Microstomus pacificus*; and a mix of other flatfishes. Combined (not including halibut) their landings totaled about 373 million pounds per year in the late 1990s.

Rock sole are well-camouflaged bottom-dwelling fish that lie in wait to attack prey that pass close by.

Halibut

Alaska's largest flatfish species, the highly prized Pacific halibut, *Hippoglossus stenolepis,* lives on the ocean floor. It is the only flatfish species not considered to be a groundfish in terms of management and markets.

Like most other fish, halibut begin life in an upright position with an eye on each side of the head. But a few days after

A left-eyed starry flounder.

Rock sole camouflaged on seafloor.

Pacific halibut.

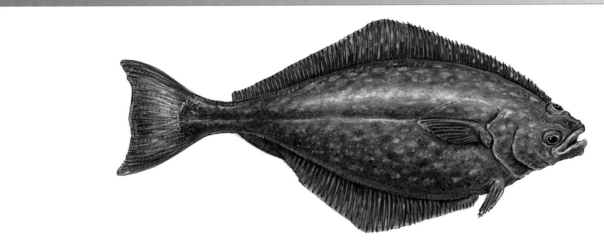

Pacific Halibut
Hippoglossus stenolepis

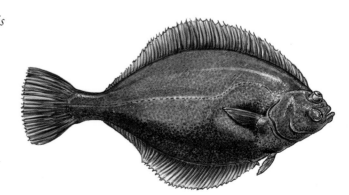

Starry Flounder
Platichthys stellatus

Yellowfin Sole
Pleuronectes asper

hatching when they are about an inch long, an extraordinary transformation begins: their body form changes and over several months the left eye moves over the snout so that both eyes are on the right side of the fish's head, and the upright fish gradually flops over onto its left side. The eyed side that faces up turns a mottled grey-brown color to camouflage it on the seafloor. The bottom side turns white.

Most halibut are "right-eyed," also known as "right-handed," with both eyes on the right side of their body (dextral). About one in every 20,000 halibut is left-eyed, or left-handed (sinestral).

Halibut are caught by longline gear, and average around 33 to 47 pounds. But many weigh over 200 pounds, and some exceed 500 pounds. Almost all halibut larger than 100 pounds are females.

In the late 1990s, Alaska's annual commercial landings averaged 58 million pounds. Halibut are sold in supermarkets and seafood stores, and the tender, pure-white fillets are popular menu items in upscale restaurants. This species also supports a big sport fishery.

Shellfish

Crabs, shrimp, clams, scallops, and other shellfish comprise another major component of Alaska's commercial fishery. Crab alone account for about 20 to 25 percent of the total value of Alaska's fisheries products.

STEPHEN JEWETT

Adult male king crabs scuttle across the seafloor on their spring migration to mate with females in shallow water near Kodiak Island. The tendency of crabs to aggregate makes pot fishing effective. The scent of bait and the presence of crabs already in the pot attract still more crabs. King crabs actually pile up against the outside of crab pots that are already full.

Ten species of crabs are harvested in Alaska: snow (opilio), red king, golden king, blue king, scarlet king, Dungeness, Tanner (bairdi), grooved Tanner, triangle Tanner, and Korean hair crabs. Snow, king, Dungeness, and Tanner (bairdi) crabs make up most of the catch. Snow crabs, often referred to by fishermen as opies, are the most plentiful and valuable by virtue of their large numbers, primarily found in the Bering Sea.

King crab eating a sea urchin.

Tanner crab.

Dungeness crab.

Red King Crab
Paralithodes camtschaticus

Golden King Crab
Lithodes aequispinus

Snow Crab
Chionoecetes opilio

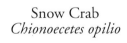

Tanner Crab
Chionoecetes bairdi

Dungeness Crab
Cancer magister

King Crabs

Three species of king crabs are commercially important in Alaska waters—red, blue, and golden (also known as brown). King crabs are caught in every coastal area of the state, from Southeast Alaska to Norton Sound. The biggest landings come from the Aleutians and eastern Bering Sea.

King crabs are mainly caught in offshore waters by larger (90- to 150-foot) boats equipped with box-shaped traps, called pots, constructed of steel frames covered with synthetic net mesh. King crabs generally segregate themselves according to age and sex.

Red king crab, *Paralithodes camtschaticus,* was the bonanza fishery of the early 1980s, with landings reaching 185 million pounds before plummeting to 15 million pounds in 1988. It still is the most abundant and commercially important king crab. In the late 1990s, the total king crab harvest averaged 18 million pounds per year as king crab stocks began to show a slow recovery. But biologists say catch levels may never again reach those of the boom years. Cause of the king crab "crash" has never been determined. But biologists see cycles of sea temperatures and abundance of predator species as corresponding to crab abundance.

Dungeness Crabs

The Dungeness crab, *Cancer magister,* is a small (2-2.5 pounds) crab with a maximum shell size of about seven inches. They inhabit bays and nearshore waters from Southcentral Alaska south to central California. The sweet, delicately flavored Dungeness is a favorite of many

KURT BYERS

Dungeness crabs mate while concealed in seaweed on a Sitka beach. Males mate only with female crabs that have just molted (shed their old exoskeleton). The female crab stores the sperm until her eggs are fully developed. The eggs are fertilized when the female extrudes them under her abdomen where she carries them until hatching. A large female can carry 2.5 million eggs.

people, as well as sea otters. This crab also is available to sport and subsistence crabbers because it inhabits shallow waters in many accessible locations.

The Dungeness is taken in cylindrical pots. Most Dungeness fishing in Alaska occurs in Southeast Alaska, near Yakutat, in Prince William Sound, and around Kodiak Island. Annual landings in the late 1990s averaged 4.7 million pounds, dropping to 2.74 million pounds in 2001.

Nomadic Herds of the Seafloor

A sight rarely seen by humans, juvenile red king crabs form a pod composed of several thousand males and females in waters near Kodiak Island. Female Tanner crabs also form pods, but only during ovulation. Little is known about podding behavior of snow crabs.

When king crabs are about one-and-a-half to two years old they emerge from rock crevices, kelp patches, and other protective niches where they have lived since they settled in as larvae.

About the size of a quarter, the young crabs wander slowly across the mud or sand bottom searching for food. They eventually join up in small groups with other crabs the same size. Different groups merge, forming increasingly larger aggregations which build into large balls of crabs, called "pods." Pods made up of thousands of crabs can reach 12 feet in length.

Podding is thought to be equivalent to herding in terrestrial animals. It probably provides protection from predators. King crab pods can be found year-round, but are most common in winter. Each pod is made up of both males and females. They usually disband at night to feed or change location. Pods are most common when population densities are high. When abundance is low, king crabs tend to be less highly aggregated, and pods become scarce.

King crabs continue to live in pods until they become sexually mature, at 5 to 7 years of age. After that, they separate by sex, but may still form loose-knit aggregations. However, old habits die hard as the older, larger crabs still tend to gather in groups—but now in scattered circular layers of males or females, two deep in the center.

Roaming gaggles of males migrate together from deep to shallow water in the spring, where they meet up with aggregations of 2,000 to 6,000 females. Then the mating begins.

Snow Crabs

Snow crab, *Chionoecetes opilio,* is Alaska's most abundant crab species. Snow crabs are found in the northern Gulf of Alaska and the Bering Sea. Alaska snow crabs support a fishery that has rapidly grown since the 1960s to an annual harvest in the late 1990s of 100 to 300 million pounds, worth about $140 million a year to fishermen. However, in 1999, the snow crab population crashed, prompting fishery managers to severely curtail the fishery.

Snow crabs are caught in stacking conical pots usually baited with chopped herring, or in large box-shaped pots like those used for king crabs. They are harvested nearshore by small boats and offshore in the Bering Sea by large boats.

Tanner Crabs

Tanner crabs, *Chionoecetes bairdi,* look very much like snow crabs and indeed, both species are often referred to as Tanners, even by some fisheries scientists. Tanners are found from Oregon north to the Aleutian Islands and southeastern Bering Sea. Adult Tanners weigh about twice as much as adult snow crabs, with a body about the size of a Dungeness crab. Tanners are much less numerous than snow crabs. The highest concentrations are found in Western Alaska waters, which yielded about 5.2 million pounds a year in the 1990s.

STEVEN REILLY

A spot shrimp creeps along a rocky Southeast Alaska seafloor.

Shrimp

Alaska waters support two relatively modest but stable shrimp fisheries—a beam trawl fishery in Southeast Alaska, which takes mostly pink shrimp, *Pandalus borealis,* and a smaller pot fishery that takes the larger-sized (known as prawns) spot shrimp, *Pandalus platyceros*; coonstriped shrimp, *Pandalus hypsinotus*; and sidestriped shrimp, *Pandalopsis dispar* in Southeast and Southcentral Alaska.

Spot and coonstriped shrimp are associated with rock piles, coral, and debris-covered bottoms. Pink, sidestriped, and humpbacks typically occur over muddy bottoms. Pink shrimp inhabit the widest depth range (60-800 feet) while humpies and coonstriped usually inhabit shallower waters (18-1,200 feet). Most shrimp start life as males, then after spawning one or more times, they turn into females, a phenomenon called protandrous hermaphroditism.

Measuring a spot shrimp.

Spot shrimp.

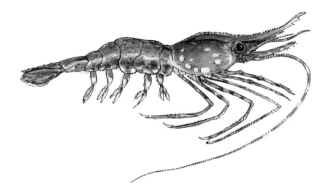

Coonstriped Shrimp
Pandalus hypsinotus

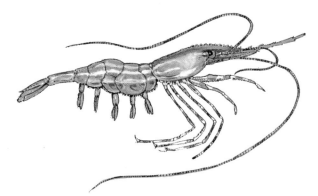

Sidestriped Shrimp
Pandalopsis dispar

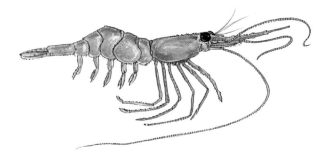

Pink Shrimp
Pandalus borealis

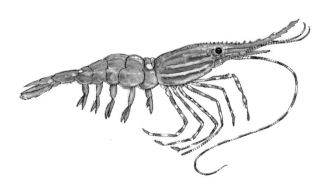

Spot Shrimp
Pandalus platyceros

Between 1960 and the early 1980s, a large trawl fishery existed in the Kodiak/Chignik/Alaska Peninsula region. But since 1985, a decrease in the stocks has reduced it to a modest fishery for a small number of pink and coonstriped shrimp. At first overfishing was blamed for the decline, but oceanographic research has since indicated that climate change played a big role.

The pink shrimp trawl harvest, which until the 1990s occurred in the Kodiak region and Southeast and Southcentral Alaska, historically made up about 80 percent of Alaska's trawl-caught shrimp harvest. The pink shrimp fishery peaked in 1976 at 129 million pounds. In the early 1980s, annual pink shrimp landings declined to less than two million pounds, and trawl fisheries for shrimp were closed in the Kodiak and Alaska Peninsula region.

By the year 2000, pink shrimp harvests recovered a bit to level out at about 3 million pounds a year, mostly from Southeast Alaska. As of 2000, the Southeast Alaska beam trawl fishery was the state's only limited entry trawl fishery for shrimp. While pink shrimp remain the primary target of beam trawl fishermen, as prices have risen for prawns, shrimp fishermen have begun to target the larger and more valuable coonstriped shrimp

The pot fishery for spot and coonstriped shrimp is conducted in the rocky nearshore waters of Southeast Alaska. This fishery yields between a half million and one million pounds of prawns per year, worth up to five dollars per pound to fishermen. Spot shrimp make up about 90 percent of the harvest.

ALASKA SEA GRANT

Often larger than those found in the Pacific Northwest, razor clams in Alaska can grow to 11 inches in length and live to be 15 years old, due to cold water and resulting slow growth rates. A large, active foot enables them to move rapidly up and down in their burrow and retreat quickly when disturbed.

Clams and Scallops

Alaska waters support a small clam industry. Prior to World War II more clam harvesting occurred. But concerns about paralytic shellfish poisoning (PSP) resulted in restrictions on the sale of Alaska clams for human consumption, and the industry has never recovered its former vitality.

Probably best known for their commercial value are Pacific razor clams, *Siliqua patula*, a tasty clam that averages about seven inches in length. Razor clams are harvested primarily from sandy

Weathervane scallop.

Geoduck.

Razor clams.

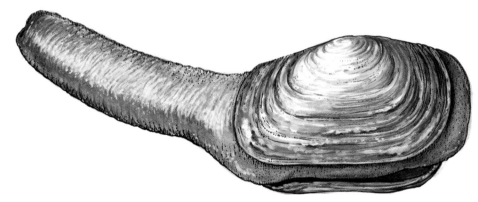

Pacific Geoduck
Panopea abrupta

Weathervane (Giant Pacific) Scallop
Patinopecten caurinus

Pacific Razor Clam
Siliqua patula

beaches and nearshore areas in Prince William Sound and Cook Inlet.

A sizable resource of Arctic surf clams, *Mactromeris polynyma*, lies below the shallow waters of the eastern Bering Sea, and at one time exploratory clam dredging was conducted there. At that time, Bering Sea clams were more expensive to harvest than their Atlantic counterparts. Interest periodically re-emerges for developing a clam fishery in the Bering Sea. But the issue is clouded by the fact that the clams are an important food for the Bering Sea's abundant and protected walrus population.

A small fishery in Southeast Alaska harvests geoducks (pronounced "gooey duck"), *Panopea abrupta*, a large clam found from shallow intertidal areas out to waters 300 feet deep or more. The largest geoducks exceed four feet in length when stretched out and weigh more than 10 pounds.

BRADLEY STEVENS

Weathervane scallops grow in large aggregations called beds. They can swim in bursts of up to 20 feet to escape predators, by squirting out jets of water. Unique among bivalves, scallops have many jewel-like eyes just inside their shell margins which detect motion and changes in light intensity.

In a technique known as hookah diving, divers expose geoducks on the seafloor by using a high-pressure water jet to blow away overlying sand or mud. See the Boats and Gear chapter for details on this technique.

The necks are skinned and sold to buyers in Japan, or are sliced and sold in seafood restaurants as "king clam." Other meaty parts of the geoduck also are marketed.

Several species of scallops occur in Alaska waters, but only one—the weathervane scallop, *Patinopecten caurinus*—supports a commercial fishery. Scallops are bivalve mollusks, like clams, and live on the sea bottom in depths of as much as 900 feet, although most are found in 120-360 feet of water. They prefer mud, clay, sand, or gravel bottoms.

Weathervane scallops, also known as giant Pacific scallops, mature in four years at a four-inch shell width. Scallops feed primarily on invertebrates, and may live up to 30 years.

Weathervane scallops occur in commercial concentrations along the coast of the Gulf of Alaska as far west as the Aleutians, in the Bering Sea, in Southeast Alaska, and in some embayments such as Cook Inlet and Prince William Sound.

The Invisible Killer

Although Alaska has abundant wild clam and mussel resources, it doesn't have much of a clam and mussel fishery. One important reason is the potential for paralytic shellfish poisoning (PSP).

PSP is a sometimes fatal condition that results from eating certain bivalve mollusks containing a toxin produced by a tiny single-cell alga of the genus Alexandrium. *The microscopic dinoflagellate is normally present in small and harmless quantities. But when stimulated by an abundance of nutrients and sunlight, they reproduce rapidly, creating a condition known as a "bloom." Filter-feeding bivalves such as clams and oysters concentrate the dinoflagellates and their byproducts in their tissues, including a poisonous compound known as saxitoxin.*

A person who eats shellfish laced with saxitoxin may experience numbness or tingling about the mouth and tongue, weakness, dizziness, nausea, and shortness of breath, leading to respiratory paralysis. Chances for survival are greatly enhanced if symptoms are quickly recognized, appropriate first aid applied, and the victim is rapidly transported to advanced medical care.

Although Alexandrium *often blooms with other algae whose cells stain the waters and cause the famous "red tide," deadly PSP can be present without a visible change in water color or other signs. And although blooms usually occur in early summer, some species of bivalves retain the toxin in their flesh for up to two years. So there is no safe time to eat them.*

Before they are allowed on the market, commercially harvested bivalves by law must be tested for PSP. It's a cumbersome process that includes shipping samples to a laboratory where technicians inject a slurry of clam extract into live mice. If the mouse dies, the fisherman is notified that he cannot market the batch of bivalves from which the sample came. At this writing there is no practical way to test bivalves taken in non-commercial (recreational and subsistence) harvests.

Scientists are working on promising new methods to test for PSP contamination. When they succeed, new opportunities for harvesting and marketing clams on a larger scale may emerge.

A family digs for clams on a Sitka beach under the watchful eye of a Marine Advisory agent from the University of Alaska Fairbanks.

Blue mussels.

The Alaska weathervane scallop fishery began in 1968 and peaked in 1978 at 1.85 million pounds shucked meat weight. Shucked meat, composed of the main adductor muscle, is the only part used. It is about seven percent of round weight. Since 1978 catches have ranged between a quarter million and 1.8 million pounds, with landed value of $1 million to $7 million. Top producing areas are the Bering Sea and the Kodiak, Kamishak Bay, and Yakutat areas.

ART SUTCH

An octopus hunts down sea urchins in a Southeast Alaska nearshore habitat. Octopus feed on urchins, shrimp, fish, bivalves, and crabs. Octopus are food for seals, sea lions, sea otters, birds, and fish.

Other Shellfish

In addition to the "big three" (crab, shrimp, and bivalves) commercial fishermen take several other shellfish species when they can find a market for them.

A few octopus are caught in specially designed octopus pots, in shrimp and crab pots, and in trawls. A small amount of Alaska octopus is sold as food, especially in Asian markets. Most is eviscerated, frozen, and sold as bait to longliners, who value it for the toughness which allows it to stay on the hook longer than other baits.

Sea cucumbers are warty, rubbery, bottom-dwelling creatures valued for the rows of delicate white meat found inside their unattractive bodies. A commercial dive fishery has developed in Southeast Alaska and around Kodiak Island, with the product going to markets on the West Coast and in Asia. Dried sea cucumber is a staple in Chinese cuisine.

Sea urchins are spiny echinoderms, bottom-dwellers that graze on kelp and other algae. The sea urchin is prized for the sweet, orange-colored, delicately flavored gonadal material, sometimes called roe, or *uni* in Japanese. Uni is one of the valued components of traditional sushi. Divers harvest urchins at various locations in Southeast Alaska and Cook Inlet, and

Sea cucumber.

Octopus.

Red sea urchin.

Giant Pacific Octopus
Octopus dofleini

Sea Cucumber
Cucumaria

Sea Urchin
Strongylocentrotus

Red sea urchins are most common in Southeast Alaska and make up most of the Alaska urchin harvest.

Green sea urchins are common in the Aleutian Islands.

Pinto abalone are no longer harvested because of high predation by sea otters. In addition to its meat, abalone shells were prized as a raw material for jewelry.

there has been exploratory effort around Kodiak Island.

The pinto abalone is a small northern cousin of the California abalone. It clings to wave-swept rocks in the intertidal and subtidal zones of ocean coastlines of Southeast Alaska. The peak season was 1979-1980, when fishermen hauled in about 378,000 pounds of abalone. But in the 1980s, an expanding sea otter population consumed most of the abalone resource and there has been no commercial harvest for years.

Other Fisheries

On certain tides in early spring, armies of kelpers wend their way across certain rocky Western Alaska beaches, picking wild popweed kelp covered with herring eggs. This product goes to the roe-on-kelp market, although at a dollar a pound it has low value compared to the other herring roe harvests. Still, an aggressive picker can make $100 an hour, creeping along on hands and knees, yanking handfuls of fish eggs and kelp off the rocks. But the fishery only lasts for one or two tides each year.

Commercial fishing also occurs in a few freshwater bodies around the state. Anadromous (hatched in freshwater, reared in salt water, returning to freshwater to spawn) smelts are taken commercially in some rivers. A few arctic char are caught and sold by arctic commercial fishermen. Whitefish, burbot, and sheefish, residents of arctic and subarctic waters, support small commercial fisheries. Nearly all commercial freshwater fisheries occur in isolated regions of Western Alaska. Local people, mainly Yupik and Inupiaq Eskimos, catch and consume most of these fish.

Underwater Deforestation

STEPHEN JEWETT

Remnants of kelp stalks on the seafloor in the Aleutians near Amchitka Island show the effects of voracious grazing by sea urchins. Sea urchins usually live in cracks in rocks and other sheltered niches which provide protection from predators, and eat algae that drift into their vicinity. But when drifting algae and/or predators are scarce, urchins will venture from their hiding places and feast on nearby seaweeds.

Underwater forests of kelp are highly productive environments and provide important habitat for many fish and other sea creatures. Sea urchins eat algae, and are especially partial to kelp. When not enough urchin predators—sea otters, sea stars, octopus, and fish—are present, urchins can devour kelp forests. The loss has an effect on the overall productivity of local marine ecosystems.

Sea otters eat a lot of urchins. So when a sea otter population declines for whatever reason, urchin populations usually grow. As the urchin population grows, kelp forests usually shrink due to overgrazing by the urchins. These grazing forays can destroy up to thousands of square feet of kelp forests.

Around 2000, sea otter populations plunged in the Aleutians, possibly due to predation by killer whales. As a consquence, urchins multiplied and devoured kelp forests.

A situation in Southeast Alaska illustrates a different sequence of events. After the Marine Mammal Protection Act went into effect, sea otters were re-introduced in Southeast Alaska in the 1980s-1990s. While that pleased a lot of people, it didn't make fishermen very happy. When otters were sparse, crabs, urchins, and bivalves were plentiful. Commercial fishing for crabs thrived, along with fisheries for urchins and pinto abalone.

When sea otters came back in force, crabs and bivalves—favorite foods of sea otters—greatly declined in number, along with commercial harvest of those species. Little was known about the abundance of urchins before or during the rebound of the sea otter population, so effects of sea otter predation on urchins in Southeast Alaska has not been well-documented.

It's an intriguing thought, but it is doubtful resource managers will ever use commercial harvest of urchins as an ecosystem management tactic to stimulate kelp bed growth in Alaska or anywhere else.

Suffice to say the killer whale/otter/urchin/kelp story is an excellent example of change that can occur when just one component of an ecosystem is added or removed, an issue that is central to many debates about the effects of commercial fishing.

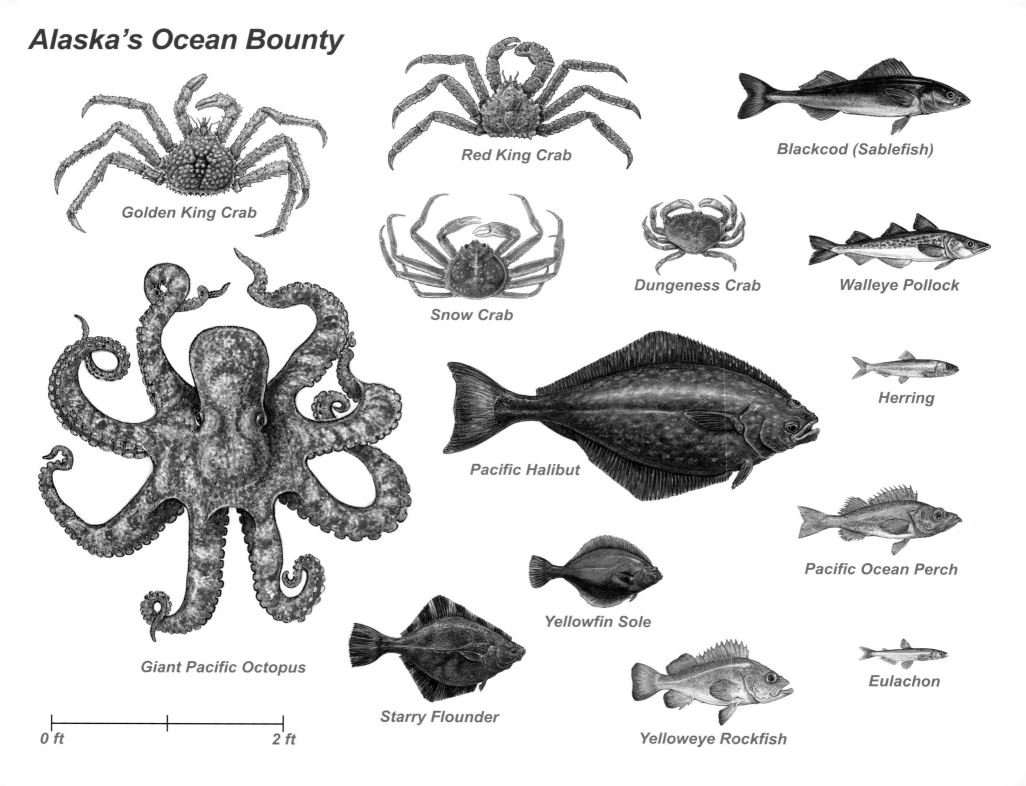

Alaska's Ocean Bounty

Golden King Crab

Red King Crab

Blackcod (Sablefish)

Snow Crab

Dungeness Crab

Walleye Pollock

Giant Pacific Octopus

Pacific Halibut

Herring

Yellowfin Sole

Pacific Ocean Perch

Starry Flounder

Yelloweye Rockfish

Eulachon

0 ft   2 ft

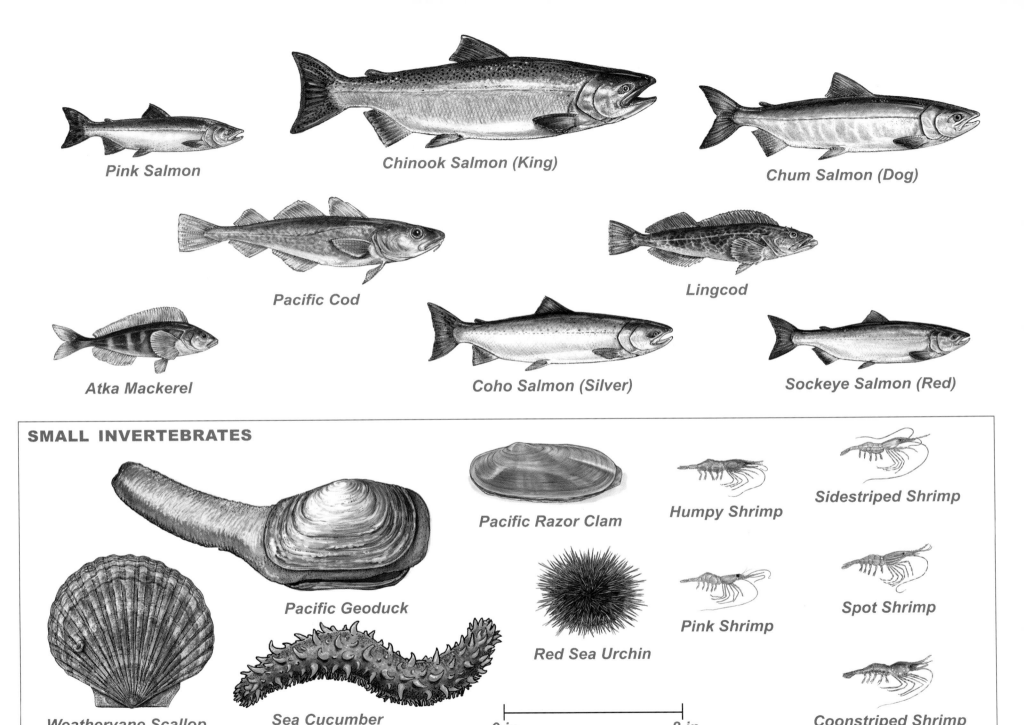

Pink Salmon

Chinook Salmon (King)

Chum Salmon (Dog)

Pacific Cod

Lingcod

Atka Mackerel

Coho Salmon (Silver)

Sockeye Salmon (Red)

SMALL INVERTEBRATES

Pacific Razor Clam

Humpy Shrimp

Sidestriped Shrimp

Pacific Geoduck

Red Sea Urchin

Pink Shrimp

Spot Shrimp

Weathervane Scallop

Sea Cucumber

0 in   8 in

Coonstriped Shrimp

Some of the best Alaska seafood is shipped immediately to fine restaurants and seafood markets, such as this one, Pike's Outdoor Market in Seattle, Washington.

DAVID BRENNER

PRODUCT FLOW

Commercial fishermen make their livings by providing the raw materials that eventually become valuable seafood products for consumers. Fishing is like farming, mining, and logging in the sense that catching the fish is only the first step in a process that results in a finished product. Fish from the sea has become part of a worldwide commodity market, affected by advances in technology, fluctuations in exchange rates, trends in contemporary culture, and all the vagaries of modern commerce.

Some seafoods are highly processed. For example, pollock caught in the Gulf of Alaska or Bering Sea is used primarily to make a paste called *surimi*, or the similar *kamaboko*. *Surimi* is shaped and textured to make imitation crab legs or shrimp tails by adding flavorings, colorings, and fillers. These products are retailed under various trade names across the country and around the world.

Pollock also is filleted and frozen, and made into blocks that are used to make fish sticks and portions for sale in supermarkets and fast food restaurants. Some of the pollock catch is sold as headed and gutted fish.

At the other extreme is fresh salmon or halibut, which are cleaned, chilled, and flown to restaurants or seafood counters within a couple of days of harvest. Some Alaska seafoods, like quillback rockfish, are not even gutted before shipping, because the market prefers whole fish. And now shippers are beginning to refine and use techniques that allow long distance shipment of live fish.

This chapter provides a general overview of how Alaska seafood goes from the fisherman's deck to your dinner table.

From Deck to Dinner Table

Whether or not extensive processing is involved, numerous steps are required to get seafood from the sea to the platter. The fact that most of Alaska's fishery products are exported to foreign markets adds to the process. Let's follow the product flow of a typical Alaska seafood item headed for Japan: sockeye salmon.

Harvest

Drifting on the deep blue, wind-chopped waters of Southeast Alaska, a 32-foot fiberglass gillnetter is retrieving its net. Crew at the stern of the vessel watch the nylon multifilament web as it comes over the roller at the transom and is wound onto a reel behind them. As a fish entangled in the web comes over the roller, the person operating the reel stops it and grabs the fish, a bright sockeye salmon of about six pounds. It is worked free of the net and placed into one of the several iced fish holds in the hull of the boat. When the holds are full, or when the fishing period ends, the skipper heads for a tender.

The tender is a 70-foot wooden vessel, its aft section filled with large tanks containing refrigerated seawater. The gillnetter's crew puts the sockeyes into a large box on a scale on board the tender. The salmon are weighed and placed into chilled seawater tanks. Within minutes the core temperature of the fish drops to about 31°F. Then the tender weighs anchor and steams to a shoreside processing plant.

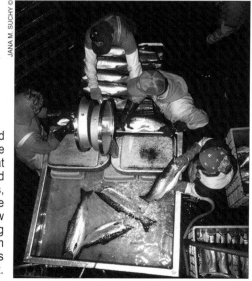

JANA M. SUCHY©

Headed and gutted salmon are moved off the "slime line." To ensure that seafood is processed under sanitary conditions, processors nationwide are required by law to follow strict sanitation monitoring procedures, a system called Hazard Analysis Critical Control Point.

Processing, Storage, Transport

At the plant the salmon are lifted in net bags or pumped into a large container which is hoisted onto the dock. Workers sort the fish by species and quality, place them in totes, and wheel them to the "slime line." There, at a long table served by a conveyor and a row of cold-water spray nozzles, workers behead, eviscerate, trim, and wash each fish.

Egg skeins are removed from the other entrails and go to a separate room where Japanese technicians supervise plant workers who wash, brine, and pack them in specially designed wood or plastic boxes destined for Japan. In some cases, the crews work the skeins on special screens to separate the eggs to make red caviar.

From this point the sockeyes could go any of several routes. Most are destined for the voracious

Supply and Demand

Salmon at Pike's Outdoor Market in Seattle, Washington.

The prices Alaska fishermen receive for their catches to a large degree reflect the economic principle of supply and demand, modified by some conditions peculiar to the fishing industry. When there is more fish (supply) landed by the producers than the consumers want (demand), prices usually fall. When the supply is smaller than consumer demand, prices usually rise. But many complex and unpredictable factors can affect both supply and demand.

Fishermen in most cases don't sell directly to consumers; rather, they sell into a chain of middlemen, starting with the processor, who sells to an importer, who sells to a wholesale distributor, and so on, through possibly a re-processor, another distributor, to a retailer or restaurant.

While ultimately the consumer expresses preferences through purchase of the final product, the middlemen can modify or even misinterpret consumer demand at several steps in the process.

For example, in 1988 growing consumer demand in Japan for salmon caused buyers to react (some say, overreact) by bidding up the ex-vessel price to record and unsustainable levels. The very next year the market's perception that oil from the Exxon Valdez oil spill might taint Alaska salmon caused buyers to lower prices paid to processors by nearly half, and did so before consumers even had a say in the matter.

Fish prices are influenced by a number of factors. They include the level of harvests of the same species in other countries (such as Canada and Russia) or of competing species. Domestic and foreign production of farmed fish, the state of the economy in consuming nations, foreign currency exchange rates, and changes in consumer nation demographics and cultural preferences affect prices, too. Leftover inventories from previous years also affect prices. High profile marketing, as well as special promotions, can stimulate demand.

Japanese market, although in recent years demand for sockeyes has increased in the United States. Some of the best and brightest might go straight to the airport for a flight to Seattle or other major city, to appear the next day on the counters of fish markets and on the menus of upscale restaurants. Others go to a huge walk-in blast freezer where they are frozen solid as cordwood and glazed with a sugar syrup to prevent dehydration and oxidation. And depending on market demand, some may be filleted before freezing. Then they are packed in labeled cardboard boxes and stored in freezer vans. Many of the fish may go to the canning line and end up in "one pound tall" or "half pound" cans.

The sockeyes remain in the freezer vans until the vans are full, then go by barge to Seattle where they are held in huge frozen-storage facilities. After that, they are transported by container ship to Japan.

At the dock or airport in Japan, the sockeyes will be inspected at the customs house and then picked up by a drayage firm. They may be taken to a wholesale market like Tokyo's sprawling Tsukiji Central Wholesale Market, or they may be held in an importer's frozen storage facility. In either case they will be examined by sharp-eyed middlemen, purchased, then trucked to reprocessors who will salt, slice, and package them for display. Other middlemen will then sell and distribute them to restaurants, supermarkets, department stores, or retail seafood stores.

KURT BYERS

Freezer vans filled with pollock, Pacific cod, flatfishes, and surimi are ready for loading onto a container ship at Kodiak.

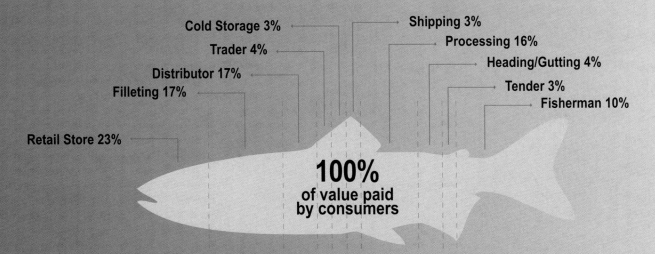

Cold Storage 3%

Shipping 3%

Trader 4%

Processing 16%

Distributor 17%

Heading/Gutting 4%

Filleting 17%

Tender 3%

Fisherman 10%

Retail Store 23%

100%
of value paid
by consumers

How Price Is Set for Salmon

Processors decide how much money to pay fishermen based on estimates of consumer demand and how many fish government scientists project will be harvested. Prices vary greatly from year to year, and even within each year. But for illustration purposes, let's talk about chum salmon and say that the price paid to fishermen (ex-vessel price) is 40 cents per pound.

Fisherman
A fisherman delivers his chum salmon to a tender and gets paid 40 cents per pound.

Tender
The tender delivers the fish to the processor and gets paid 5 cents per pound, in accordance with a contract they set up earlier.

Processor
The processor removes the roe, which is worth about 15 to 25 cents per pound of fish round weight. Head and guts are removed, which account for about 25 percent of the fish's original weight. The remaining useful flesh is worth about 73 cents per pound. The headed and gutted (H&G) fish is frozen and stored in semitrailer freezer vans in Alaska. When the vans are full, the processor transports the vans via barge, container ship, or truck to cold storage facilities, most in the Pacific Northwest.

Trader/Broker
Traders buy large quantities of the frozen H&G chum salmon from a processor's cold storage holdings for $1.63 per pound. They add about a 10 percent markup for their profit when they sell the fish to distributors. Brokers are sales agents for processors and do not physically take possession of the fish.

Distributor
The distributor buys frozen H&G fish from a wholesale trader/broker for $1.79 and adds a markup of about 20 percent. Distributors typically own cold storage facilities and a fleet of trucks. They often add value to the fish by filleting them or cutting the H&G fish into steaks. In this example, the retailer does the filleting.

Retailer
The retailer buys frozen H&G fish from a distributor for $2.15 per pound, and fillets it. Filleting is labor-intensive and adds an expense of about 90 cents per pound. That brings the cost up to about $3.05 per pound. From there, the final markup to the consumer varies greatly, from 15 percent when fish are put on sale, up to 50 percent. Average markup is about 30 percent, which in this case brings the retail price to about $4.00 per pound. The retailer's profit often is diminished when they do not sell all their fish.

Consumer
The consumer pays $4.00 per pound for chum salmon fillets at a supermarket or other retail store. Bon appetit!

MARION OWEN

Canned salmon to

TONY LARA

Roe prepared

GUY HOPPEN

13462

To processor

Unloaded and sorted

Delivered to tender

Selected for canning

Whole fish ready for shipping

Selected for freezing

ART SUTCH

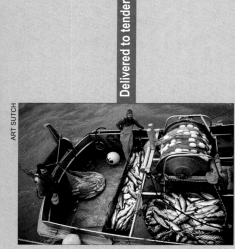

DECK

MARION OWEN

MARION OWEN

MARION OWEN

Canned fish ready for shipping

DINNER TABLE

ASMI

KURT BYERS

Roe ready for shipping

KURT BYERS

KURT BYERS

U.S. grocery stores

Japan

NATALIE FOBES

H&G frozen fish ready for shipping

U.S. fish market

MANDY MERKLEIN

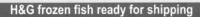

Pricing

Along this journey from gillnet to platter the fish may pass through as many as a dozen sets of hands, each of which will charge a fee or mark up the price a few cents per pound. The sockeyes for which the fisherman is paid 80 cents a pound at the tender will fetch close to eight dollars a pound in the Japanese supermarket, and considerably more if it appears in thin delicate slices in an upscale Japanese restaurant.

Nobody makes an exorbitant profit. The processor's price per pound is perhaps double what was paid to the fisherman, but the processor is discarding head and entrails, and has considerable cost in labor and overhead. The retailer adds on 15 to 50 percent at the last step, but has to cover heavy overhead expenses and losses from unsold product. Everyone in between adds five percent, ten percent, a few cents per pound—whatever it takes to get the product where it is going and provide a small return to the one supplying the service. Profits are made on volume—nearly a hundred million pounds of Alaska sockeyes might go to Japan in a bumper year.

Different Fish for Different Palates

Markets differ with each product. Blackcod largely is destined for Japan and follows a route similar to that of salmon. Much of the blackcod is smoked once it arrives in Japan.

Halibut goes mainly to the U.S. market where most is sold fresh, head off, or filleted. But some is frozen and cut into steaks with a bandsaw. Halibut rarely is smoked or canned.

This is milt from Pacific cod which will be sent to Japan where it is used as a topping for sushi dishes. Japanese buyers pay about $12.00 a pound for P-cod milt.

Soon after harvest, most crab (snow, king, and Dungeness) are cooked in the shell, either whole or in sections. The meat may be extracted later by processors outside Alaska, but most crab legs are sold as-is all over the world. Some crab is sold live.

Groundfish, including sole, cod, and pollock, are caught by trawlers and quickly delivered to nearby shore-based or floating processors where they are filleted before freezing. Fillets are packed between sheets of waxed paper in boxes known as "shatterpacks." Some fillets are "flash" frozen and packaged individually, producing a product known as "individually

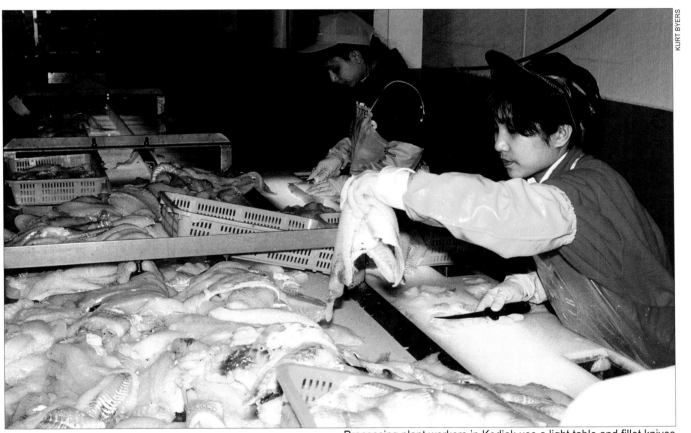

Processing plant workers in Kodiak use a light table and fillet knives
to trim Pacific cod fillets which will be boxed and frozen.

quick-frozen" (IQF) portions. Also, there is a growing seasonal market for fresh cod and flatfish fillets, which are airfreighted to buyers around the United States and abroad.

Factory trawlers process their own catch, mostly pollock and cod. Much of the pollock is made into surimi, which in turn is used to manufacture imitation crab legs and other seafood "analogs." Pollock and cod also are processed into fillets or blocks, which go to fast-food restaurant chains in the United States, and to reprocessors who make them into portioned products such as fish sticks or frozen fillets you find in the local grocery store.

Most herring are frozen and shipped to Asia, where the roe is extracted and the carcasses sold as bait, aquaculture feed, or human food. The roe is washed, brined, bleached, sorted, boxed, and sold as a high-priced delicacy to be given as gifts during Japanese holidays.

KURT BYERS

Fresh Alaska sidestriped shrimp are intermittently available from October through January. Frozen sidestriped shrimp are available in limited quantities year-round. The ex-vessel value is about $1.00 per pound.

Shrimp are the single largest component of U.S. seafood consumption. Alaska's shrimp catches are divided between two distinct markets.

The smaller pink shrimp, now mostly caught by beam trawlers in Southeast Alaska and formerly by bottom trawlers in the Gulf of Alaska and Bering Sea, are quickly absorbed into the gigantic "salad shrimp" market. They are used in a variety of prepared products as well as sold as individual peeled shrimp for shrimp cocktails and salads.

The larger but less numerous spot shrimp and their rarer cousins, sidestriped and coonstriped shrimp, are mostly pot-caught and commonly marketed as "prawns." Most spot shrimp are frozen on board catcher boats for direct sale to domestic markets, especially seafood counters and Asian restaurants in U.S. West Coast cities. Small quantities are exported to Japan.

Alaska waters yield a remarkable variety of fisheries products which accounts for about half of the entire seafood production of the United States.

Salmon take several product forms. Canned salmon is sold mainly in the South and the Upper Midwest of the United States and in Western Europe—especially the United Kingdom—and in Australia.

King and larger coho salmon go fresh or frozen to smokers in Japan, Europe, and the eastern United States, where they are processed into lox and other smoked products.

Most pinks end up in cans, but some are frozen or flown out fresh to appear as value-priced supermarket specials, and increasingly pinks also are filleted.

Many chums are kippered or made into other smoked products, and some years a percentage of the catch goes to fill the huge Japanese demand. Fresh chum fillets and dressed fish also are popular in many U.S. markets.

Fresh salmon steaks and fillets that appear in seafood market display cases and as center-of-the-plate items in upscale restaurants may be Alaska kings, sockeyes, cohos, or chums. They compete with farmed salmon from the United States, Canada, Europe, or Chile.

KURT BYERS

Not all Alaska seafood enters the big-time distribution chain. Among other direct marketing tactics, owners of this family operation in Wrangell sell their freshly canned sockeye directly to tourists who pass by on the street outside.

Market niches rapidly change and new products constantly emerge. The advent of fresh farm-reared salmon has caused a major shift in market demand, both for the restaurant trade and among European smokers. Norway, Chile, Scotland, Australia, and Canada are major farmed salmon exporters. Wild fish from Russia also are competing in the markets.

Alaska processors have lost some of their oldest markets in Europe and Japan to farmed salmon, but some of those wayward customers are returning to wild fish because of its superior color and flavor. Some processors are making Alaska fish more consumer-friendly by offering boneless fillets and portions.

The availability of farmed salmon has encouraged many people to try salmon. With better marketing and a wider array of products, Alaska seafood processors have taken advantage of this larger pool of consumers. Marketers of Alaska salmon now emphasize the healthful quality and convenience of fresh-frozen salmon from the ocean, and are tempting consumers with new products such as microwaveable fish entrees and fast-food salmon nuggets.

On average, each American consumes
about 15 pounds of seafood per year.

During an intense halibut opener, this vessel nearly sank and lost its entire catch of 10,000 pounds of halibut valued at about $13,000.

FISHERIES MANAGEMENT

It became evident early in the twentieth century that Alaska's fishing industry could catch more fish than nature could produce. Back then, salmon fisheries management consisted of simply restricting total fishing effort through measures like Sunday closures and limits on the length of nets and vessels. Despite the good intentions, salmon stocks began a long and steady decline.

Eventually the State of Alaska developed a system of salmon management based on the concept of "harvestable surplus." This means that biologists calculate how many fish are required to make it to the spawning grounds to perpetuate the stock, and fishermen are allowed to catch only the part of the run that is surplus to the required "escapement." With effective in-season run strength monitoring, the harvestable surplus approach is effective at maintaining productive fisheries and even at rebuilding depleted salmon stocks. Herring fisheries are similarly managed. For the system to work, however, biologists must be able to count fish, visually or electronically, as they approach the spawning grounds.

Other species require different approaches, and most other fisheries are managed on the basis of predetermined quotas. Ideally, quotas are based on the size of the "standing stock biomass" of the target species. In other words, biologists use test fisheries and sophisticated statistical methods to estimate the amount of each of the target species in the sea, and set total allowable catch (TAC) quotas based on those estimates.

State Management

Until Alaska statehood in 1959, the federal government managed Alaska's fisheries, offshore and in nearshore waters. The commonly held, although not necessarily correct, view was that the federal management effort was directed at maximizing the profits of Seattle-based processing companies, even to the detriment of the resource. The fight for statehood was based in part on the demand by Alaskans that they have a greater say in the management of their fisheries. One of the first acts of the newly created state legislature was to ban salmon traps, the hated symbol of outside corporate dominance of the fisheries.

UAF RASMUSON LIBRARY ARCHIVES

This fence-like structure in nearshore water traversed by migrating salmon funneled salmon into a corral-like trap.

Soon the Alaska Department of Fish and Game (ADFG) and the state Board of Fisheries and Board of Game were established, which ushered in a new emphasis on resource management.

ADFG's guiding mandate is to preserve the resource. Commercial, recreational, and subsistence fishermen are allowed to harvest only an amount of fish that is surplus to the needs of each stock to maintain itself at its maximum productivity.

Effort limitation is the main mechanism for preventing over-harvest. Limitation is accomplished through a combination of time and area closures, gear and vessel restrictions, and ultimately, license limitation.

The fisheries board system has taken the policy-making function from the bureaucrats and put it into the hands of citizens appointed by the governor—individuals recognized as being knowledgeable in the fisheries and understanding of the needs of the industry and non-commercial users of the resource. The Board of Fisheries considers proposals for regulation changes submitted by ADFG and by the public. Local advisory committees provide input, too.

Under ADFG and Board of Fisheries management, salmon landings have rebounded from historic lows of about 22 million fish in the mid-1970s to a record of 217 million in 1995 (see the History chapter for more background). Management has become increasingly sophisticated, relying on sonar counting devices, extensive computer analysis, and other management tools.

Workers haul up salmon from a fish trap
sometime in the early to mid-1900s.

Federal Management

While the ADFG–Board of Fisheries system manages most fisheries conducted in state waters, fisheries conducted in the Exclusive Economic Zone (EEZ) from three to 200 miles from shore are largely under federal management. They include virtually all of the groundfish fisheries. Some other fisheries are under a joint state-federal management arrangement, and some are under international management.

The North Pacific Fishery Management Council (NPFMC) is one of eight regional fisheries management councils nationwide established by the 1976 Fisheries Conservation and Management Act. The council makes recommendations to the U.S. Secretary of Commerce on how the harvest of fish and shellfish is to be carried out.

Composed mostly of regional appointees from the commercial fishing and processing industries, and conservation interests, the group meets five times a year to write and amend fishery management plans, determine quotas, and allocate resources among users.

After the Secretary of Commerce considers and approves recommendations from the NPFMC, those policies are carried out by the National Marine Fisheries Service (NMFS), also known as NOAA Fisheries, a federal agency in the National Oceanic and Atmospheric Administration of the U.S. Department of Commerce. The U.S. Coast Guard also has enforcement responsibilities for harvest and safety regulations.

TONY LARA

A crewman flushes pollock from a trawl net codend. The Alaska pollock fishery is the world's largest single species fishery. Management recommendations are made by the North Pacific Fishery Management Council, and approved by the U.S. Secretary of Commerce.

State and Federal Management of Alaska Fisheries

Management of Alaska commercial fisheries is complicated. Depending on location, some species are managed by either federal or state agencies, and some are managed internationally through treaties with other countries. Following is the general management responsibility by species.

State Managed

Chinook salmon	*Oncorhynchus tshawytscha*
Sockeye salmon	*Oncorhynchus nerka*
Coho salmon	*Oncorhynchus kisutch*
Pink salmon	*Oncorhynchus gorbuscha*
Chum salmon	*Oncorhynchus keta*
Pacific herring	*Clupea pallasii*
Rockfishes*	*Sebastes & Sebastolobus* sp.
Lingcod	*Ophiodon elongatus*
Snow crab	*Chionoecetes opilio*
Tanner crab	*Chionoecetes bairdi*
Grooved Tanner crab	*Chionoecetes tanneri*
Triangle Tanner crab	*Chionoecetes angulatus*
Golden king crab	*Lithodes aequispinus*
Red king crab	*Paralithodes camtschaticus*
Blue king crab	*Paralithodes platypus*
Scarlet king crab	*Lithodes couesi*
Dungeness crab	*Cancer magister*
Korean hair crab	*Erimacrus isenbeckii*
Weathervane scallop	*Patinopecten caurinus*
Snails	*Neptunea* sp.
Giant Pacific octopus	*Octopus dofleini*
Sea cucumber	*Parastichopus californicus*
Green sea urchin	*Strongylocentrotus droebachiensis*
Spot shrimp	*Pandalus platyceros*
Northern (pink) shrimp	*Pandalus borealis*
Coonstriped shrimp	*Pandalus hypsinotus*
Sidestriped shrimp	*Pandalopsis dispar*
Littleneck clam	*Protothaca staminea*
Butter clam	*Saxidomus gigantea*
Pacific razor clam	*Siliqua patula*
Pacific geoduck	*Panopea abrupta*

Federally Managed

Arrowtooth flounder	*Atheresthes stomias*
Dover sole	*Microstomus pacificus*
Flathead sole	*Hippoglossoides elassodon*
Rock sole	*Pleuronectes bilineatus*
Yellowfin sole	*Pleuronectes asper*
Pacific cod	*Gadus macrocephalus*
Pacific halibut**	*Hippoglossus stenolepis*
Walleye pollock	*Theragra chalcogramma*
Sablefish (Blackcod)	*Anoplopoma fimbria*
Skates	

* Frequently harvested in both state and federally managed fisheries.
** Managed jointly by the United States and Canada.

IFQs and CDQs

As competition for the valuable resources in the federal EEZ became intense, fishermen and managers looked for ways to assure orderly and sustainable harvests. The limited entry system which works well for salmon and herring was not appropriate for the offshore groundfish and halibut fisheries. So NPFMC explored other options.

In the mid-1990s, after about ten years of study, debate, and testimony from stakeholders, NPFMC adopted a system known as Individual Fishery Quotas (IFQs). This program gave each vessel owner who had a history of making landings in a particular fishery a quota share of the catch. A fisherman's quota is approximately equivalent to his or her average yearly landings calculated over several previous recent years. IFQs can be sold if a fisherman wants to get out of a fishery for any reason. And that is the only way someone new can get into an IFQ fishery.

The IFQ system promotes greater safety, lower operating costs, less waste of non-target species, and better prices to the fishermen by eliminating "derby" openings and extending fishing time. It has been used successfully in Canada, Australia, New Zealand, Iceland, and elsewhere, but only rarely in the United States.

In the early 1990s, the controversial plan was opposed by a vocal segment of the fleet because of limits it places on the ability of fishermen to move from one fishery to another. However, NPFMC adopted the plan for halibut and blackcod, and indicated that other species would be added in time.

IFQs have had a revolutionary effect on the Alaska halibut fishery.

A fisherman gaffs a nice blackcod (sablefish) in the Gulf of Alaska. The blackcod and halibut fisheries came under Individual Fishery Quota management in the late 1990s.

ART SUTCH

Alaska Annual Fisheries Harvest Openings

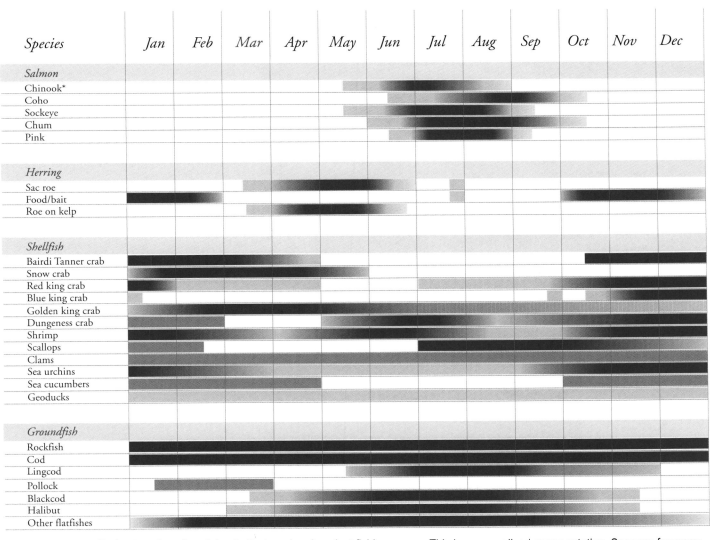

Darkest portion of each bar indicates when heaviest fishing occurs. This is a generalized representation. Seasons for many species vary depending on management region. Some fisheries close whenever a target or bycatch species quota is taken.

*Chinook trolling occurs year-round in Southeast Alaska.

Before IFQs, halibut fishermen competed in short, intense, and sometimes dangerous derby-style free-for-all seasons which lasted only a day or two, often in harsh, life-threatening weather. The photo on page 128 shows the crew of a halibut longliner during one of the hectic openers continuing to fish as their boat takes on water and comes perilously close to sinking. Soon after the photo was taken the captain astutely ordered his crew to throw overboard everything that wasn't tied down—including 10,000 pounds of halibut. The boat and crew were saved, to fish another day.

A resident of Savoonga on St. Lawrence Island suspends a big halibut harvested in that village's Community Development Quota fishery.

Now that kind of fishing intensity is gone because the quota system gives halibut fishermen the option to fish at their own discretion over an eight-month season until they catch their assigned amount of fish.

The IFQ program has been so successful for the commercial halibut fishery that in 2000 NPFMC voted to extend it to the halibut sport charter fishery, and after that to the king and snow crab fisheries. The crab plan extends quota rights to processors who take delivery of the catch, which has stirred a lot of high energy debate.

In another controversial decision, NPFMC established the Community Development Quota Program (CDQ) in the early 1990s. The setup allocates a certain percentage of the Bering Sea and Aleutian pollock, halibut, blackcod, Atka mackerel, Pacific cod, and crab harvest to several groups of cash-strapped communities, each located within 50 miles of the Bering Sea in Western Alaska.

Part of the complicated plan is a provision which initially set aside a quota of 7.5 percent of the Bering Sea pollock and cod catch for the exclusive use of residents in some 65 Bering Sea

State Managed Commercial Fishing Regions

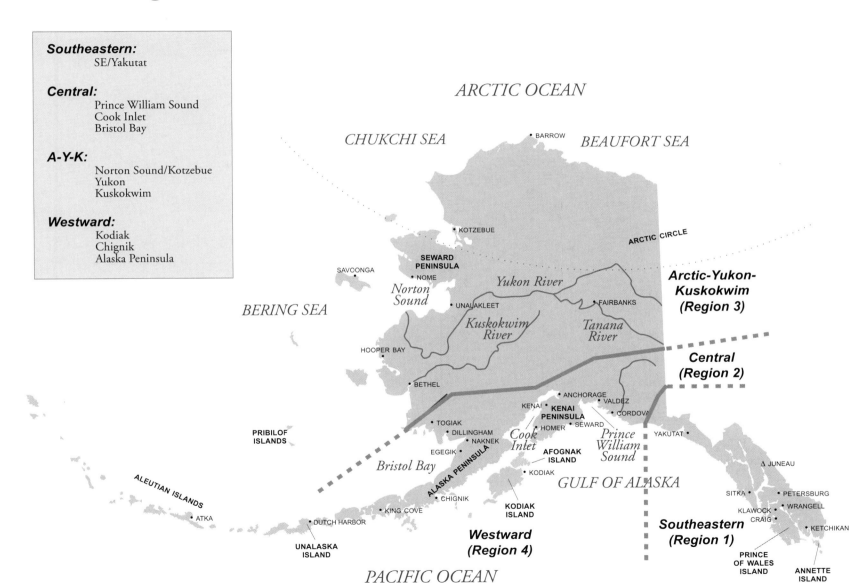

Southeastern:
SE/Yakutat

Central:
Prince William Sound
Cook Inlet
Bristol Bay

A-Y-K:
Norton Sound/Kotzebue
Yukon
Kuskokwim

Westward:
Kodiak
Chignik
Alaska Peninsula

ARCTIC OCEAN

CHUKCHI SEA

BARROW

BEAUFORT SEA

KOTZEBUE

ARCTIC CIRCLE

SEWARD PENINSULA

SAVOONGA

NOME

Yukon River

Arctic-Yukon-Kuskokwim (Region 3)

Norton Sound

UNALAKLEET

FAIRBANKS

BERING SEA

Kuskokwim River

Tanana River

HOOPER BAY

Central (Region 2)

BETHEL

ANCHORAGE

KENAI • KENAI PENINSULA

VALDEZ

CORDOVA

TOGIAK

HOMER • SEWARD

Prince William Sound

YAKUTAT

PRIBILOF ISLANDS

DILLINGHAM

Cook Inlet

NAKNEK

EGEGIK

AFOGNAK ISLAND

JUNEAU

Bristol Bay

ALASKA PENINSULA

KODIAK

GULF OF ALASKA

ALEUTIAN ISLANDS

CHIGNIK

KODIAK ISLAND

SITKA •

PETERSBURG

ATKA

KING COVE

KLAWOCK • WRANGELL

CRAIG

KETCHIKAN

DUTCH HARBOR

Westward (Region 4)

Southeastern (Region 1)

UNALASKA ISLAND

PACIFIC OCEAN

PRINCE OF WALES ISLAND

ANNETTE ISLAND

coastal communities. This move allowed residents in those largely subsistence-oriented towns and villages to participate in the burgeoning pollock fishery off their shores and receive some of the income generated by what was then a billion-dollar enterprise.

However, it was clear from the outset that fishermen from those tiny villages were unlikely to harvest all the fish they were entitled to. So the communities—through regional nonprofit corporations—formed joint ventures with the large factory trawler companies and shore-based processors to share the income from fish sales.

Under the CDQ program, the large fishing companies are training and employing residents of CDQ villages in Western Alaska. And the CDQ groups use their earnings to build infrastructure such as seafood processing plants, provide training in seafood industry jobs for local residents, and offer other economic help.

According to State of Alaska figures, in the decade following institution of CDQs in 1992, the effort generated about 9,000 jobs for residents of Western Alaska with wages totalling about $60 million.

KURT BYERS

A resident walks up the beach in Hooper Bay, a village of approximately 215 households on the Bering Sea coast. This and more than 60 other cash-strapped Bering Sea coastal villages augment their economies with assets channeled to them via the Community Development Quota Program.

Counting Heads

Biologists from the Alaska Department of Fish and Game collect salmon on the Yukon River.

When it comes to counting fish and shellfish, scientists do what fishermen do—they go fishing.

Systematically fishing key areas of the ocean gives scientists a better idea of the relative fish abundance, size, sex composition, and other factors essential to managing fisheries. They use the same gear as fishermen—trawl nets and longlines, or pots in the case of crab species. But scientists don't try to catch a lot of fish or crab. They catch samples. Then they use complex statistical formulas and historical data to estimate the number or volume (biomass) of fish or shellfish.

In a few cases, other techniques are used to estimate biomass. For example, herring form dense schools close to the water surface when spawning time arrives. The schools are easily seen from above, so biologists take a bird's eye view from airplanes and come up with estimates that way. Another technique is sonar. Sonar surveys allow researchers to gather an electronic signature of a school of fish. Mathematical formulas are then used to calculate the size of the stock.

There is at least one fish group that is usually counted one at a time—salmon—and they often are counted twice. They're counted first as they head out to sea after hatching in freshwater, and again when the fish return as adults. One unique experimental technique mates old technology with new. Scientists have attached video cameras to fish wheels to get a constant, 24-hour, 7 days a week count of salmon as they come through the fish wheel. That data is then used to estimate the total number of fish migrating up the river.

Scientists have tried to count fish for decades. But too often, estimates prove to be off the mark—so there is still a lot to be learned about the seemingly simple task of counting fish.

Federal Law Conserves Pollock

America's biggest fishery, Alaska pollock, became the subject of special legislation in 1998 with passage of the American Fisheries Act. Championed by the Alaska congressional delegation, the highly controversial law closed loopholes in Coast Guard policy that had allowed foreign-owned and -rebuilt vessels to operate off Alaska's shores. The law reduced the number of factory trawlers in the fishery and restricted new entry, changed ownership nationality requirements, and paved the way for creation of producer cooperatives.

After that, companies still operating in the fishery formed the Pollock Conservation Co-op. Participants devised a way to share the allowable catch and eliminate the race to land the quota. Purpose of the co-op was to make the fishery safer, reduce capital and operating costs, minimize bycatch, and optimize utilization.

International Management

Some fishery resources are distributed across national boundaries and their management lies with international agencies. Best known in terms of its influence in Alaska is the International Pacific Halibut Commission (IPHC), which sets halibut harvest quotas for the North Pacific and Bering Sea. Canada and the United States jointly participate in the IPHC, whose annual meetings alternate between U.S. and Canadian locations. The IPHC has done an outstanding job since the 1930s of maintaining the vitality of the halibut fishery and of orchestrating its recovery when stocks were depleted by foreign trawlers.

Another international organization

MANDY MERKLEIN

Walleye pollock fill a trawl net. While not the most commercially valuable marine species per pound—shrimp, salmon, crab, and other species are much more valuable—pollock is the nation's second most valuable commercial species because of the harvest's extraodinary volume.

What's the Catch?

Living with and working alongside fishermen, fisheries observers estimate total catch, identify more than 85 species of fish and shellfish, and count, measure, sex, and weigh them.

To make wise management decisions, fisheries managers need to have a good idea of how many commercially important fish and shellfish live in the ocean and monitor how many are caught by fishermen.

State, federal, and university scientists doggedly pursue the difficult task of stock assessment—that is, calculating the volume of each species of interest in the sea.

To assist this effort, some Alaska fisheries include a program whereby specially trained people, called fisheries observers, ride aboard fishing vessels to monitor and report to shoreside managers what's being caught—bycatch and target species. The managers use the information to decide when it's time to open and close fishing seasons.

Every year, over 300 men and women from all over the United States spend 30 to 90 days aboard fishing vessels in the Bering Sea and Gulf of Alaska collecting data on what fishermen catch. They work up to 12 hours a day, 7 days a week aboard trawlers, longliners, and pot vessels that are fishing for groundfish, crab, or scallops. In 2000, a trial observer program for marine mammal encounters began in one of Alaska's salmon fisheries.

These people are part of the largest and most comprehensive observer program in the world, administered by the U.S. Department of Commerce National Marine Fisheries Service and the Alaska Department of Fish and Game.

Fishing vessels in Alaska are required by law to carry observers if they are longer than 65 feet and fish in federal waters. Fishermen are also required to pay for observers, which at this writing is often over $200 a day per observer—and sometimes more than one observer is aboard. Part of that money goes to contractors who recruit and place observers on vessels. While observers are not always the most welcome people onboard, the industry has accepted their presence and cost, realizing that observer coverage is one of the key components to maintaining healthy fish stocks.

influential in Alaska fisheries is the Pacific Salmon Commission (PSC), a joint U.S.-Canada agency. A long history of U.S. and Canadian fishermen catching salmon produced by each other's national waters demonstrated the need for international management. The main goal is to reduce interception of each other's salmon and rebuild the stocks.

PSC determines the percentage of trans-boundary salmon U.S. and Canadian fishermen can harvest each year. The fishery management agencies of each country make specific regulations to ensure compliance. In only a few years of existence, PSC efforts spurred a notable increase in stocks of chinook salmon in Southeast Alaska, British Columbia, and the Pacific Northwest states. But the PSC has experienced difficulty in attempting to satisfy all parties.

Yet another organization was created in 1992 to bring together the United States, Canada, Japan, and Russia to conserve high seas stocks of salmon and related species. The North Pacific Anadromous Fish Commission oversees the international ban on high seas salmon fishing and coordinates research on migratory stocks in the North Pacific and Bering Sea. The convention that created the commission was the death knell for Asian high seas driftnet fleets, which for decades had decimated North American and Russian salmon stocks and killed large numbers of non-target fish, seabirds, and marine mammals.

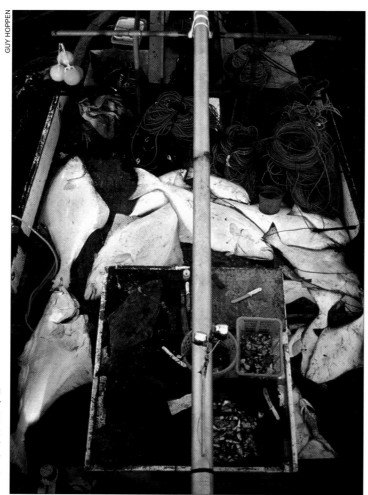

GUY HOPPEN

Halibut just pulled from the water are ready for immediate dressing and refrigeration. The North American halibut fishery is jointly managed by the United States and Canada.

Where Have the Sea Lions Gone?

Steller sea lions (named after German naturalist Georg Wilhelm Steller, who accompanied Danish explorer Vitus Bering to Alaska) look for an easy meal from a trawler. To the consternation of fishermen, these imposing animals, weighing up to a ton, sometimes haul out and bask in the sun on the sterns of fishing boats.

Alaska's Steller sea lions are in trouble. The eastern Gulf of Alaska stock (inhabiting the waters of Southeast Alaska) is listed as Threatened under the federal Endangered Species Act, although their numbers have increased a little in recent years. But the western stock which inhabits the western Gulf of Alaska, Aleutian Islands, and southern Bering Sea, has suffered a population drop of about 80 percent since the 1970s to only about 40,000 animals by 2000, prompting an Endangered classification. In 2002, scientists saw the first increase in the western stock since the plunge started in the 1970s.

Scientists have considered a variety of possible causes for the decline, including incidental takes in the commercial fisheries, subsistence harvests, disease, predation by killer whales, migration out of the region, and lack of food.

Research indicates that juvenile sea lions experience the greatest mortality, and that food stress may be a primary cause of their problem. The decline of the western stock has paralleled the development of the offshore trawl fishery in the same region. The main target of that fishery is Alaska pollock, which also happens to be the single largest component of the sea lion diet in the Gulf of Alaska and eastern Aleutians. So it seems logical that there could be a cause-and-effect relationship. But the ocean ecosystem is not that simple.

Pollock are not as nutritious for sea lions as much fattier forage species that sea lions used to depend on, such as sand lance, smelt, capelin, and herring—fish which themselves have declined over the past few decades. Research has shown that captive sea lions actually lose weight when provided a diet consisting only of pollock.

Pollock stocks grew enormously starting in the 1970s, apparently due to changing ocean temperatures, while stocks of more nutritious forage fishes decreased. So it also could be that climate change has played a major role in the demise of Steller sea lions in Western Alaska.

Chum salmon are prepared for drying at Nelson Island.

SUBSISTENCE FISHING

Most of Alaska's Native people lived—and continue to live—on the coast and along major rivers. Fish have always been their key to survival. Before Alaska had a commercial fishery, indeed, before there was a cash economy of any kind, Alaska's Native people caught fish to eat. For at least 9,000 years, Native people trapped, speared, netted, and hooked fish for local trade and personal consumption. Although Alaska is relatively rich in game, the pursuit of fast-moving large animals was a much riskier proposition than harvesting fish that literally swam to the people.

Naturally, in a matter as important as securing food, human ingenuity flourished. Using only the raw materials at hand, Native Alaskans devised remarkably clever and efficient means of catching fish.

These traditional food-gathering activities, which continue today, are called "subsistence" fishing. This tradition has become the focus of some of the most bitter political wrangling in state history due to conflicting views on issues such as equal access and preferential rights of rural people to food resources. Subsistence issues often are associated with Native people, but in current practice all rural Alaska residents—Native and non-Native—have subsistence rights, and urban Natives are treated like other city-dwellers.

While lawyers, politicians, activists, and bureaucrats struggle to sort out the best legal approach to managing subsistence fisheries, thousands of rural people, Native and non-Native alike, quietly go about their business of catching and preserving the fish they need to feed their families.

Traps

Fish traps are one of the best examples of Native ingenuity. The basic concept of a fish trap is to put a fixed structure in a location where migrating or feeding fish will swim into it and not find their way out.

Small, basket-like woven traps are still used in the Yukon-Kuskokwim Delta area by the Yup'ik Eskimo people to catch blackfish, a small fish found in lakes and ponds. The trap is baited with meat and lowered from a skiff into the lake or positioned through a hole in the ice. Fish approach the trap, which is resting on the lake bottom, and follow its periphery until they locate the funnel-like entrance. Inside, the trap opens into a large chamber. The blackfish lose track of the opening and cannot find their way out. After a day or so, the fisherman retrieves the trap.

This is a model of a typical Native fish trap. Attracted by bait in the narrow end, fish enter the trap through the open end and funnel themselves through the outermost cone into the rear cone where they become trapped.

Equally effective (and simpler) traps were used in tidal areas. One elegant technique took advantage of the fact that when the fish arrive at their small spawning streams, they typically swim into a stream mouth, mill around for awhile, then drift out into the sea again. The salmon repeat this behavior until their eggs ripen and their instincts indicate it is time to proceed upstream to the spawning grounds.

Native fishermen would line up a row of large stones, extending them outward from one bank in a curve or hook pattern. The stones were placed so that on the incoming tide salmon could easily swim over the rock barrier. But when the tide dropped and the fish were ready to drift back to the bay, their exit would be blocked. When the tide dropped enough, the people had only to walk over and pick up the fish.

Other traps were made with stakes driven into the river bottom to form a kind of weir or pen. Archaeologists have located many of these intertidal traps at river mouths. With the advent of commercial fisheries, traps evolved into large floating affairs many yards square with long leads—like fences—extending out from the shoreline to divert fish into them.

Most kinds of fish traps are now banned in Alaska, but in a few places their use continues. The Tsimshian Indians at Annette Island near Ketchikan are allowed four large, floating salmon traps for commercial harvest in the waters adjacent to the island. However, due to changing economic conditions they have not used them in recent years. Smaller traps for smelt, blackfish, and other freshwater fish still are used in various parts of Alaska.

Skiffs used for subsistence and commercial
fishing line the beach at Hooper Bay.

Spears and Hooks

If the trap was the most efficient gear, either the spear or hook-and-line was the least efficient. Still, under the right conditions, each had its place.

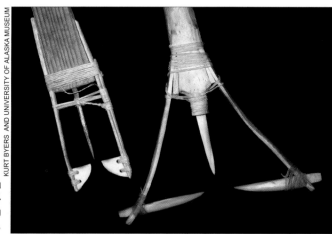

Two examples of the business end of a leister. These fish spears were about 4 to 6 feet long. Points were made from whale bone, antlers, wood, and steel.

The flounder-shaped lure on this jig rig is made of baleen with two ivory eyes.

Salmon in small streams could easily be taken with a spear. So could pike and other freshwater fish in the shallow sloughs and ponds. The tool of choice is called a "leister," a spear with two barbed points.

Until introduction of the steel fishhook by Europeans, the familiar hook-shaped fishhook was unknown in Alaska. Natives used other devices for the same purpose. In Western Alaska, Eskimos jigged through the ice for cod and freshwater fish, using two-piece lures of bone, ivory or wood, bound with baleen or sinew.

In Southeast Alaska, Tlingit and Haida fishermen used an ingenious hook-like device to catch halibut. The composite hook had a carved wood section with a bone or iron barb bound to it. The lure would be attached to a line made of kelp stalks anchored with a stone, and sometimes baited with a piece of octopus. This enticing contraption would waggle in the current until an unsuspecting halibut took it. When the fish tried to spit it out, the barbs would lodge in its mouth.

In some locations rod-and-reel is a legal subsistence fishing technique, and subsistence halibut fishermen are allowed to fish short groundlines with multiple baited hooks in the same manner as commercial longliners.

Hook History

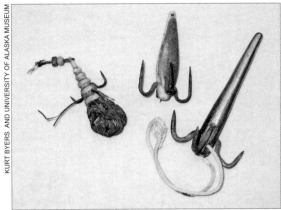

Above are three examples of hooks made by Alaska subsistence fishermen after the introduction of metal hooks.

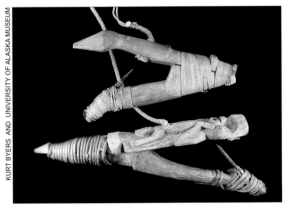

Above are two carved halibut hooks. At right, a model of how these hooks were rigged to catch halibut.

Float

Baited lure

Fish Wheels

Historians believe the first fish wheel appeared in Alaska in 1904 on the Tanana River. The technology quickly spread, and since then salmon have been harvested with fish wheels on several of Alaska's largest rivers. Today, fish wheels are used in Alaska's subsistence, commercial, and personal use fisheries. The only commercial use is on the Yukon River, where fewer than 200 permits are issued.

Operating on the same principle as a water wheel, these ingenious devices use only the river current to capture fish, with a minimum of human effort. Fish-catching baskets and paddles are alternately built onto the end of spokes. The unit is precisely positioned in the river current parallel to shore in the path of migrating salmon. Two side extensions funnel approaching salmon toward the wheel where they are scooped up by the baskets. As the wheel continues its rotation, the captured fish fall out of the baskets into a chute and slide down into a box attached to the side of the fish wheel.

To achieve best performance, many factors need to be considered. Proper siting of the wheel along the riverbank is perhaps most critical. Fish wheels must be placed at a spot where the fish migrate close to shore without being too deep or shallow for the wheel to run. A mere few feet either way can mean the difference between success and failure.

If all goes well, once set in motion fish wheels are like perpetual motion fishing machines, scooping up salmon until the run is through or season closes.

KEN KOLLEDGE

Both Native and non-Native fishermen use fish wheels for commercial fishing on the upper Yukon River system and for subsistence fishing on the Yukon, Kuskokwim, and Copper river systems. Pictured here are personal use fish wheels on the Copper River.

Preserving the Subsistence Lifestyle

A Southeast Alaska subsistence harvester gathers herring roe-on-kelp.

According to the Alaska Constitution, every resident has the right of equal access to fish and game resources. But with a significant population increase over the last two decades, the resources cannot necessarily provide for everyone—recreational, commercial, and subsistence users—who wants fish. So allocations must be made.

Under the federal Alaska National Interest Lands Conservation Act (ANILCA), subsistence priority on federal lands (most of the state) must go to rural residents. In the 1980s, Alaska went through a lengthy and painful process to determine who was a rural resident, based on the size and traditional food-gathering habits of the community, only to have the rural preference principle struck down by the Alaska Supreme Court.

Because the Alaska constitution mandates equal access to fish and game for all state residents, it is at odds with ANILCA. In the early 2000s, the state-federal impasse still was not settled. No matter how the subsistence dilemma is resolved, many Alaska residents will be left unsatisfied. But regardless of legislative and regulatory wranglings, subsistence fishing and hunting continue to support an important traditional lifestyle in Alaska's rural villages.

Nets

Native peoples have long used various kinds of nets for subsistence fishing and continue to do so today. Originally they used dipnets or small beach seines, making their web from baleen, sinew, nettles, and cedar or willow bark fibers. Now the web is synthetic (probably made in Japan), but the way it is used is the same.

Gillnets are the most commonly used subsistence fishing device in Western Alaska. In Southeast Alaska, most people use beach seines. In some areas people use long-handled dip nets to scoop up migrating salmon.

In Bristol Bay the shorelines near towns and villages are lined with clothesline-like pulley systems used to run subsistence gillnets into and out of the tide. Residents abandon their homes, offices, and work sites twice a day to check their nets, pick out the sticks and debris, and retrieve the snagged salmon that will become a central part of their winter diet.

JOHN VAN AMERONGEN, ALASKA FISHERMAN'S JOURNAL

Herring dry on a rack at Mekoryuk in Western Alaska.

Preservation

Catching fish is only half the story. Unless it is caught in weather so cold that it freezes immediately, fish is a highly perishable food. Native people have to deal with an entire year's supply of food in just a few weeks. That means the whole family pitches in with the work of fishing and fish preservation.

As soon as fish are caught, the women go to work gutting and washing. Some fish are split and dried, others cut into thin strips and smoked. Both drying and smoking removes moisture from the flesh, retarding spoilage. Fish heads and other parts sometimes are buried in the ground and allowed to ferment. Smaller fish like herring are skewered or strung on twine and air dried. All parts are used, including the eggs, which are dried and fermented.

With the advent of canning, and later, freezing, fish preservation is much easier, at least wherever these technologies can be used economically. In remote Alaska Bush communities, the drying rack and smokehouse are still the predominant systems used to preserve fish.

Chum salmon dry on a rack at Nelson Island.

Terms Used in This Book

A

Alaska Board of Fisheries: The state board of citizens appointed by the governor, who make management policy for fisheries that occur in state waters.

Alaska Department of Fish and Game (ADFG): The state agency that conducts research and implements management decisions over many of Alaska's fish and wildlife resources.

Alaska National Interest Lands Conservation Act (ANILCA): The federal law that in 1980 set aside large expanses of Alaska land in parks, preserves, and wildlife refuges.

Anadromous: A type of fish life cycle wherein the fish is hatched in freshwater, goes to sea to feed and mature, and returns to the river or lake to spawn.

Analog, seafood: A seafood product made from surimi, often in a form that looks and tastes like more expensive seafood such as crab or shrimp.

Aquaculture: Farming fish, invertebrates, or plants in freshwater. The word also is often used to refer to saltwater farming, correctly known as "mariculture."

B

Baleen: The fibrous material found in the mouths of certain species of whales, which strains out small fish and crustaceans the whale swallows as food. Native people make a variety of craft items from baleen.

Beam trawler: A specialized trawl vessel used to catch shrimp, characterized by a wood or metal beam that holds the mouth of the net open.

Benthic: On or in the floor of the ocean or other water body.

Biomass: Total estimated weight of living organisms in a given area. In this book, fish or shellfish populations.

Bivalve: Shellfish that have two shell halves, such as clams, cockles, mussels, and oysters.

Bloom: Sudden great increase in population of algae. Red tides are caused by algae blooms.

Boom: A piece of pole or pipe attached horizontally to a boat's mast and extending aft or pivoting upward and to the sides. Used as an attachment point for lights or for machinery, such as a power block.

Bottomfish: See "groundfish."

Bowpicker: A drift gillnet boat that deploys and retrieves the net over the bow (front end).

Brailer: A large, hydraulic-powered dipnet used to scoop fish from a purse seine. Also, a net or mesh bag lowered into or mounted inside a fish hold and hoisted out full of fish to speed unloading the vessel.

Brine: Water with measured amounts of salt added, used to chill or preserve fish.

Bulwark: The structure around the periphery of a boat hull which extends above deck level, forming a low wall that keeps items on deck from sliding off into the sea, and on which railings, line rollers, stabilizer poles, and other items are attached.

Bunt: The part of a seine net that becomes a bag holding captured fish as the rest of the net is "dried up," or hoisted out of the water.

Buoy line: The line that connects the anchor on the seafloor to a set of buoys on the surface. The buoys are large and brightly colored so they can be easily seen, and often have a flag attached for even greater visibility.

Bycatch: Fish and other animals caught incidentally in fishing operations but, because of species, size, condition, or other factors, are not desirable to the market.

C

Carapace: The body shell of a crab.

Chafing gear: Heavy rope matting attached to the bottom of trawl nets to protect the codend from damage as it passes over coarse seafloors.

Chilled seawater (CSW): A system for chilling the catch that uses water cooled by flake ice. Most CSW systems are set up with tubes that blow air into the ice-water mixture to prevent clumping.

Codend: The small-mesh, sock-like rear end of a trawl net where the catch is concentrated.

Community Development Quota (CDQ): A system created by the federal government and carried out by the state whereby coastal communities along the Bering Sea coast are granted a share of the total allowable catch of specified fisheries resources. Regional CDQ development groups can use the quota to fish on their own, or form partnerships with fishing companies.

Continental shelf: Shallow sea area adjacent to a continent, generally considered to be from shore out to waters 600 feet deep. It usually is the richest region of the oceans in ratio of plants and animals to water volume.

Corkline: The top line on a fishing net which floats on the surface, buoyed by styrofoam or cork floats spaced evenly along the line.

Crabber: A vessel that deploys gear ("pots" or traps) to catch crabs.

D

Demersal: Near the bottom of a water body.

Derby fishing: A situation wherein a fishery is open for a short time and participants rush to catch every fish they can during that time.

Dinoflagellate: Single-celled, microscopic, free-floating marine alga with a tail-like appendage(s) used for propulsion. One of the most basic of foods in the marine food chain.

Dipnet: A large-diameter, long-handled net that one person can use to scoop fish, usually while standing on a beach or river bank, or wading in shallow water.

Discard: Any fish or shellfish intentionally thrown away or returned to the sea dead. There are two types of discards: regulatory and economic. Regulatory discards are those species fishermen are not allowed to keep, like an under-sized chinook salmon or a halibut caught in a crab fishery. Economic discards are fish species that have little or no commercial value and are thrown overboard to save space in the fish hold for the more valuable target species.

Distant water: High seas off the shores of other nations.

Door: Heavy steel panels attached to each side of the mouth of a bottom trawl net. As the net is towed, the force of the water pushes the doors outward, and that action holds the mouth of the net open.

Dory schooner: A classic vessel style with the helm steering control located aft (to the rear), which had small, flat-bottom rowboats (dories) stacked (nested) on the decks. Hook-and-line fishing for cod and halibut was done from dories.

Drayage: Hauling a load of some commodity; in this book, seafood.

Dredge: A steel cage device that is dragged along the seafloor to scoop up scallops.

Dressed weight: The weight of a fish after it is gutted or headed and gutted.

Drift gillnet: A net held vertically in the water by a row of floats on the top and weighted line on the bottom, which entangles fish that swim into it. Drift gillnets are deployed from boats and drift along with the tide or current.

E

Echinoderm: A marine animal in the phylum *Echinodermata* which includes sea urchins, sea cucumbers, and sea stars.

Emergency position indicating radio beacon (EPIRB): An electronic transmitter which, in emergencies such as a boat sinking, can be triggered automatically or manually to send a signal to an orbiting satellite, indicating to rescuers the location of the emergency.

Escapement: The portion of a run or return of migratory anadromous fish, such as salmon, that successfully returns to spawning grounds to reproduce.

Eulachon: A small, smelt-like anadromous fish, *Thaleichthys pacificus*. Eulachon are valuable food for birds, marine mammals, and other fish, and avid recreational fishermen flock to streams to dipnet them when thousands of the fish enter creeks and rivers to spawn. Commonly known as "hooligan."

Eviscerate: To remove the guts.

Exclusive Economic Zone (EEZ): Another term for FCZ, it also includes U.S. federal control of economic resources other than fisheries, such as petroleum or seabed minerals.

Ex-vessel: The price paid to fishermen for their catch.

F

Factory trawler: A vessel that tows a net to catch bottom-dwelling or midwater fish, and processes them onboard.

Fairleads: Devices that guide lines or nets going in or out. For example, on the stern of gillnet boats are two upright rollers that guide the incoming net onto the reel. Also, the donut-shaped insulators on the end of the taglines that help spread the trolling wires from the boat are called fairleads.

Farming, salmon: Raising salmon in submerged pens, usually until the fish are mature and ready for harvest.

Fathom: A measure of length or depth, equal to six feet or approximately 1.85 meters.

Fillet: Also spelled filet, a longitudinal slice of meat taken from the side of a fish. May be boneless or with pin bones left in.

Fish meal: A product made from byproducts of fish processing—heads, viscera, frames, and undersized fish. It is used for fertilizer, animal feeds including aquaculture feed, and sometimes as a human food supplement, depending on its quality.

Fish ticket: A written record of the weight and species composition of a fisherman's catch delivered to a tender or processor. It is used by processors to determine payment to fishermen, and provides fisheries managers with a record of the volume of fish or shellfish landed.

Fish trap: Any of various devices consisting of enclosures into which fish swim and cannot get out. Salmon traps are attached to pilings driven into the sea bottom or to huge anchors, and are positioned to block nearshore fish migration routes.

Fish wheel: A device consisting of an anchored or fixed raft or scow fitted with a set of baskets that revolve like a waterwheel with the pressure of the current, scooping fish from the water. Used in some rivers for catching salmon.

Fishery Conservation and Management Act: The 1976 U.S. federal law that created the 200-mile territorial waters boundary and the regional fishery management councils. Also referred to today as the Magnuson Act and the Magnuson-Stevens Fishery Conservation and Management Act.

Fishery Conservation Zone (FCZ): The expanse of sea extending from three miles to 200 miles offshore within which the U.S. federal government exercises exclusive fishery management authority.

Flupsy: Floating upweller system. A technique used to circulate water in shellfish culture pens.

Freezer van: A shipping container, trailer, or truck that is insulated and equipped with mechanical refrigeration to keep products frozen during transport.

Frond: A leaf of a fern, palm, or algae; in this book, a kelp leaf.

G

Gaff: A tool consisting of a steel hook or spike attached to a long handle, driven into a fish's head to aid in lifting it aboard a boat.

Gangion: The short lines that connect baited hooks to groundlines used in longline fishing.

Gillnet: A net that entangles fish by the gills and fins as the fish try to swim through it.

Global positioning system (GPS): An electronic tool that processes signals from orbiting satellites to indicate one's precise geographic position.

Groundfish: Also known as bottomfish or white fish, includes bottom-dwelling and midwater-dwelling species such as flatfishes, rockfishes, Atka mackerel, pollock, blackcod, and Pacific cod.

Groundline: A long piece of heavy line to which baited hooks are attached by short lines (gangions), used in longline fishing for halibut and other groundfish.

Grow-out stage: In shellfish aquaculture, the growth period from shellfish seed planting to final market size.

Gumby suit: See "immersion suit."

Gunnel: Also spelled gunwale, the uppermost part of the hull of a vessel where the guns were mounted on a warship. It is where chocks, longline rollers, or stabilizer poles may be mounted.

Gurdy: A small winch or reel powered by a hydraulic or electric motor or by hand crank, used to retrieve trolling wire or groundline.

H

Haida: The largest Native group on southern Prince of Wales Island, Alaska, and in the Queen Charlotte Islands of British Columbia.

Hang: To "hang a net" is to build a net by lacing web to the corkline (top) and leadline (bottom).

Hanging bait: Usually a whole fish which is pinned into the interior of a crab pot to hold the attention of crabs inside the pot so they do not try to find their way out. Crabs are initially lured into the pot by another piece of aromatic bait, usually ground herring, packed inside a perforated plastic jar.

Harp: A 3-sided chute that looks like the upper half of a question mark (?). It is mounted on the stern of a boat used in the longline fisheries to pay out the gear. The gear, including the buoy lines, running line, and the ground or working line, exit the boat via the chute.

Harvestable surplus: The quantity of fish available to be caught after sufficient reserves are set aside for spawning.

Haul back: Retrieve nets or lines.

Hauler: Any of various types of devices used to retrieve nets or lines. An example is a hydraulic pot hauler, used to pull up crab pots.

High seas: Ocean waters outside the national 200-mile limit.

High seas drift nets: Gillnets, some up to 30 miles long, deployed offshore by vessels from certain foreign nations to catch a variety of species, including billfish, tuna, and squid. High seas drift gillnets are banned in parts of the North Pacific to protect migrating salmon, marine mammals, and birds.

Hold: The interior space in a vessel that contains the catch.

Hookah: A diving technique commonly used in the harvest of sea urchins, sea cucumbers, abalone, and geoducks whereby the diver, wearing a dry suit, is supplied air pumped through a hose from a tender boat. "Hookah" is the transliterated Urdu word for water pipe, used for smoking tobacco in Middle Eastern countries.

Hooligan: See "eulachon."

Hypothermia: Lowering of the body's core temperature to the point of impaired function and possibly death.

I

Immersion suit: A bright reddish-orange, full-body (including feet, hands, and head) flotation and foam-cell insulation coverall which keeps the wearer protected from extreme cold and afloat if in the water. Also known as a "survival suit" and "gumby suit."

Individual Fishery Quota (IFQ): An allocation of a specific percentage of the total annual allowable catch of a stock of fish, issued to an individual or vessel.

Individually quick-frozen portions (IQF): Portions of fish, usually boneless fillets, that are quickly frozen under extraordinarily cold temperatures, about –40°F. Quick freezing, which takes about 30 minutes, reduces formation of ice crystals in the fish tissue that form during slower freezing. Fewer ice crystals mean less breakdown of cell structure in the meat, resulting in a firmer, better quality seafood product.

International Pacific Halibut Commission: The U.S.-Canada organization that manages Pacific halibut.

Inupiaq: Inuit Eskimo culture and language, in Western Alaska from Unalakleet northward across northern Alaska, Canada, and Greenland.

J

Jig: A weighted fishing lure that is jerked up and down on the end of a line to attract fish or squid.

Joint venture: In commercial fishing, a business arrangement between two or more companies to carry out a fishing or processing enterprise. The term entered the fishing lexicon to describe arrangements that involved American catcher boats delivering to foreign processing ships during the "Americanization" of the Exclusive Economic Zone.

K

Kamaboko: A rubbery product similar to surimi which is a popular food item in Japan.

Kazunoko: Salted herring roe.

Kazunoko kombu: Kelp coated with herring roe.

Kippered salmon: A product made by hot-smoking salmon to make a flavorful but drier form of smoked fish.

L

Leadline: The weighted bottom line that holds a fishing net vertical in the water.

Leister: A spear made of various combinations of ivory, bone, antlers, and wood, traditionally used by Alaska Natives to catch fish.

Limited entry: A system of controlling fishing effort by restricting the number of permits issued for each fishery.

Limited entry permit: A permanent license issued to an individual fisherman which allows him or her to engage in a particular fishery. The permit is issued to an individual by the state on the basis of a point system, and can be sold or transferred to another individual.

Longline: A type of fishing gear made up of three distinct line segments: buoy line, running line, and working line. The working line section has baited hooks attached at regular intervals and lies on the seafloor.

Lox: Lightly salted ("mild cured"), thin-sliced salmon, a delicacy in the United States and Europe. Made with the highest quality Pacific king and Atlantic salmon.

M

Macrocystis: A type of kelp that occurs in nearshore ocean waters. In some areas, herring deposit their eggs on *Macrocystis* fronds, which can be harvested in specially designated commercial fisheries.

Mariculture: Farming fish, invertebrates, or plants in seawater. Also called aquaculture, although strictly defined, aquaculture is farming in freshwater.

Maximum sustainable yield (MSY): The greatest amount of a fish or crustacean stock that can be removed on a continuing basis without diminishing the stock's ability to replenish itself.

Mesh: The pattern (size and shape) of the web in a fish net, or the strands of twine that make the web.

Milt: Fish semen.

Mintaiko: Seasoned pollock eggs, marketed as a delicacy in Japan.

N

National Marine Fisheries Service (NMFS): The federal agency in the U.S. Department of Commerce charged with conducting research and managing fisheries within the Exclusive Economic Zone. This agency is now called NOAA Fisheries.

National Oceanic and Atmospheric Administration (NOAA): The parent agency of NMFS in the U.S. Department of Commerce. NOAA also contains the National Ocean Survey, National Weather Service, and National Sea Grant College Program.

Nearshore: Close to land, usually thought of as inside the 3-mile state waters limit.

Non-target species: Any animal—fish, bird, mammal, crustacean—caught unintentionally in a fishing harvest. See "bycatch."

Nori: Any of several species of kelp which, after being dried, are used in Japanese cuisine.

North Pacific Anadromous Fish Commission: An international body composed of the United States, Canada, Japan, and Russia which addresses issues concerning salmon on the high seas of the Pacific Ocean and Gulf of Alaska.

North Pacific Fishery Management Council (NPFMC): The federal board appointed by the U.S. Secretary of Commerce which makes management decisions on fisheries in the Fisheries Conservation Zone off the Alaska coast.

O

Ocean ranching: A salmon hatchery process whereby salmon eggs are incubated and hatched and the young salmon released to live in the ocean. When they migrate back to their release point to spawn, a portion is harvested by recreational and commercial fishermen, and the rest by the hatchery to take more eggs and milt for the next year's production.

Offshore: The region of the ocean usually thought of as beyond 3 miles from shore.

P

P-cod: Pacific cod, *Gadus macrocephalus*, also known as grey cod or true cod.

Pacific Salmon Commission: The U.S.-Canada organization that makes allocation decisions to implement provisions of the U.S.-Canada salmon treaty.

Paralytic shellfish poisoning (PSP): Human illness caused by a naturally occurring toxin produced by a species of dinoflagellate alga. People can get PSP by eating bivalve mollusks, such as clams and mussels, which have accumulated the toxin by consuming toxin-producing dinoflagellates.

Pelagic: Open ocean waters, excluding bottom and shore. Also refers to higher up in the water column or near the surface, as opposed to near the bottom (demersal).

Picking boom: A small boom on which a gear hauler, such as a crab block or a snatchblock, is hung that can be swiveled out (some types of booms are fixed in place) over the side of a boat. It allows the hauler or block to be suspended over the side so that the gear will clear the side of the boat. Also, a knuckle or articulating boom, hydraulically activated, which lifts crab pots into a boat.

Pin bones: Small unattached bones found in the flesh of some fish, such as salmon.

Pod: A mass of from several dozen up to 6,000 young Tanner or king crabs. It is most likely a defensive "herding" behavior to protect themselves from predators. Also, a group of marine mammals such as whales.

Pot: A stainless steel wire- or polyester-mesh cage-like device equipped with a trap door placed on the ocean bottom to capture shellfish and finfish lured inside by bait.

Pot launcher: Device used to lift and send pots over the side of fishing boats.

Pound: A floating enclosure used to contain herring or other fish. In one type of pound, herring deposit millions of eggs on kelp fronds tied inside the enclosure. Afterward the herring are released unharmed and the roe-on-kelp (kazunoko kombu) is harvested.

Power block: A hydraulically powered hauler, usually attached to the end of a boom above a vessel's working deck, used to pull the net into the boat.

Purse seine: See "seine."

Q

Quota: A specific amount of fish or shellfish (usually expressed in pounds) that is legally available for harvest in a set time period.

R

Raceway: A large concrete or fiberglass trough through which fresh water flows, used to contain juvenile fish while they grow in hatcheries.

Ranching, salmon: See "ocean ranching."

Raw fish tax: Revenue collected by the state based on the value of the catch when it is taken off the fishing vessel (ex-vessel value) prior to processing.

Recruit: An individual fish that is a member of a larger spawning population of the same age, or a fish that has grown large enough to be desirable for harvest.

Red tide: Seawater discolored by an algal bloom.

Refrigerated seawater (RSW): A system for chilling the catch that uses water cooled by mechanical refrigeration. May be either circulating water in a tank or a shower of refrigerated water (spray brine) inside a fish hold.

Resource allocation: The management process of determining which group of users gets what amount of the available resources.

Retort: A pressure cooker used to cook cans of salmon. The retort operates on steam and can cook up to 15,000 cans at one time, depending on size of retort and cans.

Roe: Fish eggs, either separate or clumped together in the skein.

Roller: A horizontal apparatus that assists retrieval of longline gear. It sits on the bulwark and looks like a baker's rolling pin, usually with vertical fairleads on each end which prevent the line or net from jumping off the roller. During the retrieval process, it is the first part of the boat that longline gear touches. Its function is to reduce friction and wear on the lines and on the boat as the lines come aboard.

Round weight: Total weight of a whole fish or shellfish, as opposed to "dressed weight," which is the weight of the animal after guts, or head and guts, are removed.

Running line: The hookless section of a longline groundline closest to the groundline anchor. It has no hooks so that it will not get tangled with the buoy line, which also is attached to the anchor when the anchor is being set. Also, in some setnet operations, a line strung between attachment points to which the net is clipped or tied when deployed.

S

Saxitoxin: The toxic chemical that causes paralytic shellfish poisoning in humans, produced by a dinoflagellate alga and passed on to humans when they eat shellfish contaminated with the toxin.

Scow: Large, flat-bottomed boat with square ends, often towed or pushed by tugboats. It is used to transport bulk cargo like sand or coal. A power scow is a type of fish tender driven by its own engines and equipped with crew accommodations, fish holds, etc.

Sea farming: See "mariculture."

Seine: A type of net deployed to encircle a school of fish, then pulled shut around the bottom, trapping the fish; also called a purse seine.

Set: A "set of gear" describes a group of objects—such as lines, hooks, buoys, and anchors—which, when properly connected, will catch fish. To "set gear" is to place gear in the water to fish.

Setnet: A gillnet that extends out from shore with one end anchored or fixed to the beach and the other end anchored out in the water perpendicular to the expected path of the target fish, usually salmon. Herring setnets are anchored at both ends and may be parallel, perpendicular, or oblique to the shore.

Skate: A standard length of baited groundline in a longline rig, usually 1,800 feet long.

Skein: The sac or filamentous tissue that holds large masses of fish eggs together.

Skiff: A small, open motorboat or rowboat.

Slime line: The area of a processing plant or cannery where fish are beheaded, eviscerated, and "slimed" (washed).

Slush: A combination of water and crushed ice used to chill fish or shellfish in a boat's fish hold and on to market. Also called "chilled seawater" (CSW), it differs from mechanically cooled "refrigerated seawater" (RSW).

Snap gear: Groundlines with gangions attached by stainless steel snaps. Part of longline gear.

Soak: To allow pots or baited longlines to fish. "Soak time" is the time pots or longlines are left in the water to fish.

Spotter plane: A small single-engine aircraft (or helicopter) hired by fishermen to locate schools of fish in the water or to locate vessels that are catching fish. The spotter radios the information to the vessels that have contracted for the service.

Spray brine: Refrigerated seawater sprayed over fish in a boat hold to keep the fish chilled until delivery to a processing plant.

Spread: A piece of trolling gear consisting of a snap, a two- to six-fathom length of leader, and a lure or baited hook.

Sternpicker: A drift gillnet vessel that deploys and retrieves the net over the stern (back end).

Stinger: The wand attached to the end of an air hose used by dive harvest fishermen to blow sandy bottom sediments away from and thus expose buried siphons of geoducks.

Stock: A geographically discrete population of a single species of fish or other animals.

Stuck gear: Groundlines with the gangions permanently attached, part of longline gear.

Subsistence: The traditional harvest and consumption of fish and wildlife for purposes of feeding the harvester's family and neighbors, or in some cases their domestic animals. Often associated with, but not limited to, Native peoples.

Substrate: The solid material on or in which a plant or animal lives; for example, a rocky seafloor or a sandy beach.

Surimi: A flavorless paste, usually made from pollock and other groundfishes, which can be processed into artificial crab and other seafood analogs.

Survival suit: See "immersion suit."

Sweep chain: A heavy chain attached to the leading edge of a scallop dredge, which dislodges scallops from the ocean bottom.

T

Tagline: A piece of twine or cable with attachments that allow it to guide the direction of another line, such as trolling wires.

Tallyman: Person who records the weight and species composition of a fisherman's catch when delivered to a tender or processing plant.

Tarako: Seasoned pollock roe, marketed as a delicacy in Japan.

Tattletale: A piece of twine, stretched between the tip of a trolling pole and the bow of the vessel, which jiggles to signal that a fish is caught on one of the lures.

Tender: A vessel, usually owned by or under contract to a seafood processing company, that receives fish from catcher vessels and delivers them to processing plants.

Tlingit: The largest Native group in Southeast Alaska, which extends as far north as Yakutat.

Tote: A metal or plastic box used to move or hold fish temporarily, usually in quantities of 500 to 1,200 pounds.

Transom: The flat, vertical surface that makes up the stern of most fishing vessels.

Trawler: A vessel that catches groundfish by towing an open-mouth, sock-shaped net, either along the bottom or up in the water column.

Troller: A vessel that catches salmon by towing baited hooks or artificial lures through the water, from lines weighted by lead balls suspended from long poles that extend out at a 45-degree angle from the boat's sides.

Tsukiji Wholesale Market: One of the central wholesale markets in Tokyo, most famous as the wholesale distribution point for massive volumes of fish and seafood products.

U

Uni: The gonads of sea urchins, eaten raw in sushi. Considered a delicacy in Japan and Japanese restaurants in the United States.

V

Value added: An expression that refers to processing a raw product to make it more valuable. In seafood, value added processing could include filleting, steaking, portioning, smoking, or producing more elaborate products such as microwaveable entrees complete with garnish and sauces.

W

Water column: A vertical section of a water body, from the bottom to the surface.

Web: Mono- or multi-filament nylon mesh used to make nets. Web openings are square or diamond shaped.

Weir: A structure placed in a stream that blocks or diverts passage of fish, to allow counting or catching them.

Wing: The two, large-mesh outer sides of the opening of a trawl net that funnel fish into the body of the net.

Working line: Often called the groundline, it is the part of the longline system rigged with baited hooks that lies on the seafloor and does the actual catching of the fish.

Y

Yesterday: The day when a lot of fish were where a fisherman is today.

Yup'ik: The largest Eskimo group on the Bering Sea coast, from Bristol Bay, Alaska, north to southern Norton Sound, and inland up the Nushagak, Kuskokwim, and Yukon rivers.

Z

Zero tolerance: The federal ruling that allows the Coast Guard to seize vessels on which illegal drugs have been found.

Zipper: The panel in a trawl net codend that can be released to spill the catch onto the boat's deck.

Zooplankton: Swimming or floating animals, usually microscopic, common in marine and aquatic waters. Eaten by fishes and some marine mammals, they are a primary component of the foundation of marine and aquatic food webs.

Sources and Useful References

Alaska Habitat Management Guide: Life History and Habitat Requirements of Fish and Wildlife. Juneau, Alaska: Alaska Department of Fish and Game, 1985.

Alaska Wildlife Notebook Series. Juneau, Alaska: Alaska Department of Fish and Game, 1999.

Barr, L., and Barr, N. *Under Alaskan Seas: The Shallow Water Marine Invertebrates.* Anchorage, Alaska: Alaska Northwest Publishing Company, 1983.

Blau, F., compiler. Information pamphlet on commercial fisheries, 1994-98, Kodiak, Alaska: Alaska Department of Fish and Game, Division of Commercial Fisheries Westward Region, 1999.

Charton, B. *The Facts on File Dictionary of Marine Science.* Brooklyn, New York: Facts on File, 2001.

Coughenower, D., and Blood, C. *Flat Out Facts about Halibut.* Fairbanks, Alaska: Alaska Sea Grant College Program, 1997.

Fisheries Bycatch: Consequences and Management. Fairbanks, Alaska: Alaska Sea Grant College Program, 1997.

Fisheries of the United States. Silver Spring, Maryland: U.S. Department of Commerce, National Marine Fisheries Service Current Fishery Statistics Series, 1995-2002.

Forage Fishes in Marine Ecosystems. Fairbanks, Alaska: Alaska Sea Grant College Program, 1997.

Frennette, B., McNair, M., and Savikko, H., compilers. *Catch and Production in Alaska's Commercial Fisheries, Special Publication No. 11.* Juneau, Alaska: Alaska Department of Fish and Game, 1995.

Gay, J. *Commercial Fishing in Alaska,* Vol. 24, no. 3. Anchorage, Alaska: Alaska Geographic, 1997.

Harbo, R. M. *Shells and Shellfish of the Pacific Northwest.* Madeira Park, Canada: Harbour Publishing, 1997.

Hart, J. L. *Pacific Fishes of Canada.* Ottawa, Canada: Fisheries Research Board of Canada, 1973.

Hawkins, D. M., et al. *Plant-Animal Interactions in the Marine Benthos.* New York: Oxford University Press, 1992.

High Latitude Crabs: Biology, Management, and Economics. Fairbanks, Alaska: Alaska Sea Grant College Program, 1996.

Is It Food? Addressing Marine Mammal and Seabird Declines. Fairbanks, Alaska: Alaska Sea Grant College Program, 1993.

Johnson, T., editor. *Alaska Fisheries Handbook.* Sitka, Alaska: Seatic Publishing, 1990.

Johnson, T., and Wiese, C. *Understanding Salmon Markets.* Fairbanks, Alaska: Alaska Sea Grant College Program, 1995.

Kessler, D. W. *Alaska's Saltwater Fishes and Other Sea Life: A Field Guide.* Anchorage, Alaska: Alaska Northwest Publishing Company, 1985.

Knapp, G. Inside the Roe Business. *Currents: A Journal of Salmon Market Trends.* Anchorage, Alaska: University of Alaska Anchorage. Salmon Market Information Service, March 1995.

Knapp, G. A Fisherman's Guide to Seafood Distribution. *Currents: A Journal of Salmon Market Trends.* Anchorage, Alaska: University of Alaska Anchorage. Salmon Market Information Service, March 1995.

Knapp, G. Alaska Salmon Prices: A Matter of Supply and Demand? *Currents: A Journal of Salmon Market Trends.* Anchorage, Alaska: University of Alaska Anchorage. Salmon Market Information Service, June 1996.

Kramer, D. E., et al. *Guide to Northeast Pacific Flatfishes: Families Bothidae, Cynoglossidae, and Pleuronectidae.* Fairbanks, Alaska: Alaska Sea Grant College Program, 1995.

Kramer, D. E., and O'Connell, V. M. *Guide to Northeast Pacific Rockfishes: Genera Sebastes and Sebastolobus.* Fairbanks, Alaska: Alaska Sea Grant College Program, 1995.

Kruse, G. et al., editors. *Management Strategies for Exploited Fish Populations.* Fairbanks, Alaska: Alaska Sea Grant College Program, 1993.

Larsson, A. K. *Operation and Construction of the Plumb Staff Beam Trawl.* Marine Advisory Bulletin No. 4. Fairbanks, Alaska: Alaska Sea Grant Program, 1975.

Legacy of an Oil Spill: 1999 Status Report. Anchorage, Alaska: Exxon Valdez Oil Spill Trustee Council, 1999.

Mecklenburg, C. W., Mecklenburg, T. A., and Thorsteinson, L. K. *Fishes of Alaska.* Bethesda. Maryland: American Fisheries Society, 2002.

Nybakken, J. W. *Marine Biology.* Third edition. New York: Harper Collins College Publishers, 1993.

O'Clair, R. M., and O'Clair, C. E. *Southeast Alaska's Rocky Shores: Animals.* Auke Bay, Alaska: Plant Press, 1998.

Pearcy, W. G. *Ocean Ecology of North Pacific Salmonids.* Seattle, Washington: Washington Sea Grant College Program, 1992.

Proceedings of the International Symposium on North Pacific Flatfish. Fairbanks, Alaska: Alaska Sea Grant College Program, 1995.

Restoration Update, Vol. 6, no. 2. Anchorage, Alaska: Exxon Valdez Oil Spill Trustee Council, 1999.

Ruppert, E. E., and Barnes, R. D. *Invertebrate Zoology,* Sixth Edition. Fort Worth, Texas: Saunders College Publishing, 1994.

Salmon Market Bulletin. Vol. 1, no. 7. Juneau, Alaska: McDowell Group, Salmon Market Information Service, October 1999.

Snively, G. *Exploring the Seashore in British Columbia, Washington and Oregon.* Seattle, Washington: Pacific Search Press, 1987.

Solving Bycatch: Considerations for Today and Tomorrow. Fairbanks, Alaska: Alaska Sea Grant College Program, 1996.

The Alaska Almanac. Twenty-fifth Edition. Bothell, Washington: Alaska Northwest Books, 2001.

Thompson, L., compiler. *The Bering Sea Ecosystem.* Washington, D.C.: National Academy Press, 1996.

Wallace, W. K., and Fletcher, K. M. *Understanding Fisheries Management.* Mobile, Alabama: Auburn University Marine Extension and Research Center, Sea Grant Extension, 2001.

Sunset over the Bering Sea.

Anatomy of a Public Information Project

In the late 1980s, Alaska's visitor industry began to blossom, and by the late 1990s it rivaled the seafood industry in economic importance. Much of that growth was due to the attraction of Alaska's remarkable marine and coastal resources. A key facet of Alaska's coastal region is its bustling commercial fishing communities.

In 1989, we decided it was a good time to help coastal communities focus on their local seafood industries as components in the rich mix of attractions people enjoy seeing and learning about when they visit Alaska. We looked at New England as a model of how a colorful waterfront industry could define a region in the public's perception and play an important economic role as a visitor attraction. We also realized fishing ports could serve as a backdrop for educating people about an important regional industry and lifestyle.

Our project began with an idea hatched by University of Alaska Fairbanks professor emeritus and former Homer-based Marine Advisory agent, Douglas Coughenower. Doug envisioned developing a walking tour of the harbor in Homer, anchored by an interpretive sign with a tour map and drawings of the different kinds of boats typically seen in Homer's harbor.

This book is part of a public information project that began with production and installation of interpretive signs about commercial fishing in Ketchikan, Seward, and Wrangell, Alaska.

We enlarged on Doug's concept and developed a pilot project. Local communities throughout the state were invited to work with us to develop interpretive signs with commercial fishing vessels and the fish and shellfish species typically brought to port in each town. The idea was to provide accurate and interesting dockside information, attractively displayed, to help visitors and local residents better understand the local fishing industry.

The other two components of the project were an informational brochure about the industry, custom-written for each town, and this book, intended for anyone who wanted more detailed information about Alaska's commercial fishing industry.

The towns of Ketchikan, Seward, and Wrangell responded to our invitation. Working closely with officials and fishermen in each town, we developed and placed large, colorful interpretive signs overlooking five harbors. Seward opted to include the interpretive brochure. Most of the marine species drawings seen in this book were originally commissioned for use on the signs and brochure.

We believe this document serves as a historical milepost that presents an authoritative and accurate portrait of Alaska's commercial fishing industry at a challenging time. We hope you find this book enlightening, and we welcome your comments.

—*Kurt Byers, Communications Manager/Editor, Alaska Sea Grant College Program*

About Sea Grant

Sea Grant programs sponsor scientific research and discovery about America's oceans and Great Lakes, and transfer new knowledge about those resources to people who can put it to good use. Sea Grant is a partnership and a bridge between government, academia, industry, scientists, and private citizens designed to help Americans understand and sustainably use our ocean and Great Lakes waters.

Modeled on the nation's Land Grant concept, Sea Grant was created by an Act of Congress in 1965. National headquarters are housed with the National Oceanic and Atmospheric Administration, U.S. Department of Commerce, in Silver Spring, Maryland.

Individual Sea Grant programs are located at colleges and universities in every ocean and Great Lakes coast state and Puerto Rico. The programs provide funding and other support to scientists, students, and resource specialists in more than 300 academic institutions, which generate research-based information about marine and Great Lakes resources. Outreach professionals translate and transfer this new knowledge to anyone who can use it.

This team effort of discovery and information transfer helps people make better decisions on how to use, revitalize, sustain, and enjoy our nation's priceless ocean and aquatic natural resources.

Among other subjects, nationwide Sea Grant research and outreach personnel address endangered species, aquaculture, coastal hazards, waterfront development, commercial and recreational fisheries, coastal economic development, marine safety, seafood technology and safety, coastal habitat enhancement, marine biotechnology, K-12 and college education, and other subjects critical to the long-term health of our oceans and Great Lakes.

The University of Alaska joined the Sea Grant College network in 1970. Since then, Alaska Sea Grant has sponsored university research, education, and advisory services on topics including:

- Seafood quality and product development
- Shellfish mariculture
- Commercial fishing safety
- Marine mammal and fisheries biology
- Marine ecosystem change
- K-12 and college student education
- Coastal tourism and recreation

Alaska Sea Grant's work in these and other areas benefits government decision makers, industry members, scientists and technicians, conservationists, educators and students at all levels, and the public at large. For a detailed description of Alaska Sea Grant's current research and outreach, contact Alaska Sea Grant headquarters at the University of Alaska Fairbanks, or visit www.uaf.edu/seagrant.

Want to learn more about Alaska's marine resources and fishing industry?

Here is a list of some of the fine low-cost and free Sea Grant publications that provide details about the seafood industry and the environment upon which it depends. For a catalog of all Alaska Sea Grant books, brochures, posters, and videos, contact Alaska Sea Grant, P.O. Box 755040, Fairbanks, AK 99775-5040, or call (907) 474-6707, toll-free 1-888-789-0090, fax (907) 474-6285, email fypubs@uaf.edu, or visit our convenient online bookstore at www.uaf.edu/seagrant/bookstore/. Full descriptions of these and more items are in the Web bookstore catalog.

Popular Books

Beating the Odds on Northern Waters: A Guide to Fishing Safety
Biological Field Techniques for Chionoecetes Crabs
Fish and Fisheries Elementary School Curriculum Guide
Fisheries Management for Fishermen
Fishing for Octopus: A Guide for Commercial Fishermen
Gillnet Hanging
Growing Shellfish in Alaska
Guide to Marine Mammals of Alaska
Guide to Northeast Pacific Flatfishes
Guide to Northeast Pacific Rockfishes
Marketing and Shipping Live Aquatic Products
Temperature Directed Fishing
The Bering Sea and Aleutian Islands: Region of Wonders
Who Has the Legal Right to Fish?

Posters

Alaska Wet and Wild
Alaska's Ocean Bounty
Coho Salmon Life Cycle
Common Bivalves of Alaska
Marine Mammals of Alaska

Brochures

Beachwalk
Fish-Wurst: Recipes for Sausage from Fish
Flat Out Facts about Halibut
Paralytic Shellfish Poisoning: The Alaska Problem
Rockfishes of Alaska
Seafood Safety: What Consumers Need to Know
Understanding Salmon Markets

Videos

Alaska's Inshore-Offshore Bottomfish Processing Debate
Alaska's Salmon Marketing Crisis
Alaska's Subsistence Dilemma
Beating the Odds: Onboard Emergency Drills
Halibut Dressing
Hanging a Gillnet
Marine Protected Areas in Alaska
Ocean Ranching
Sea Survival
Sharing the Sea: Alaska's CDQ Program
Steller Sea Lions: In Jeopardy
The Future of Alaska's Salmon Industry
Trouble in the Bering Sea

Science Books

Cold Water Diving for Science
Crabs in Cold Water Regions: Biology, Management, and Economics
Dynamics of the Bering Sea
Fisheries Bycatch: Consequences and Management
Fishery Stock Assessment Models
Forage Fishes in Marine Ecosystems
Is It Food? Addressing Marine Mammal and Seabird Declines
Making Profits out of Seafood Wastes
Management Strategies for Exploited Fish Populations
North Pacific Flatfish Proceedings
Our Common Shores and Our Common Challenge: Environmental
   Protection of the Pacific
Solving Bycatch: Considerations for Today and Tomorrow
Steller Sea Lion Decline: Is It Food II

Index

Page numbers in italic typeface denote photographs or illustrations